Long-Term Institutional Management of U.S. Department of Energy Legacy Waste Sites

Committee on the Remediation of Buried and Tank Wastes

Board on Radioactive Waste Management
Commission on Geosciences, Environment, and Resources

National Research Council

NATIONAL ACADEMY PRESS
Washington, D.C.

NATIONAL ACADEMY PRESS • 2101 Constitution Ave., N.W. • Washington, DC 20418

NOTICE: The project that is the subject of this report was approved by the Governing Board of the National Research Council, whose members are drawn from the councils of the National Academy of Sciences, the National Academy of Engineering, and the Institute of Medicine. The members of the committee responsible for the report were chosen for their special competences and with regard for appropriate balance.

Support for this study was provided by the U.S. Department of Energy, under Grants No. DE-FC01-94EW54069 and DE-FC01-99EW59049. All opinions, findings, conclusions, and recommendations expressed herein are those of the authors and do not necessarily reflect the views of the Department of Energy.

International Standard Book Number: 0-309-07186-0

Additional copies of this report are available from:
National Academy Press
2101 Constitution Avenue, N.W.
Box 285
Washington, DC 20055
800-624-6242
202-334-3313 (in the Washington metropolitan area)
http://www.nap.edu

Cover Image: The U.S. Department of Energy is engaged in an effort of national importance to address the legacy of environmental contamination resulting from its Cold War defense mission. Many U.S. defense complex sites are highly contaminated with radionuclides and hazardous chemicals and cannot be cleaned up with current technologies (top image). These sites must be managed to isolate and contain the waste (fence image in center), in some cases in perpetuity (hourglass image). These contaminated sites will require long-term institutional management to protect the land and the people who live "outside the fence" (bottom image). The bottom image also embodies the hope that in at least some instances the principles of long-term institutional management, diligently applied, make the need for the fence eventually to go away, as sites that are contaminated today are cleaned up in the future with new technologies.

Copyright 2000 by the National Academy of Sciences. All rights reserved.

Printed in the United States of America.

THE NATIONAL ACADEMIES

National Academy of Sciences
National Academy of Engineering
Institute of Medicine
National Research Council

The **National Academy of Sciences** is a private, nonprofit, self-perpetuating society of distinguished scholars engaged in scientific and engineering research, dedicated to the furtherance of science and technology and to their use for the general welfare. Upon the authority of the charter granted to it by the Congress in 1863, the Academy has a mandate that requires it to advise the federal government on scientific and technical matters. Dr. Bruce M. Alberts is president of the National Academy of Sciences.

The **National Academy of Engineering** was established in 1964, under the charter of the National Academy of Sciences, as a parallel organization of outstanding engineers. It is autonomous in its administration and in the selection of its members, sharing with the National Academy of Sciences the responsibility for advising the federal government. The National Academy of Engineering also sponsors engineering programs aimed at meeting national needs, encourages education and research, and recognizes the superior achievements of engineers. Dr. William A. Wulf is president of the National Academy of Engineering.

The **Institute of Medicine** was established in 1970 by the National Academy of Sciences to secure the services of eminent members of appropriate professions in the examination of policy matters pertaining to the health of the public. The Institute acts under the responsibility given to the National Academy of Sciences by its congressional charter to be an adviser to the federal government and, upon its own initiative, to identify issues of medical care, research, and education. Dr. Kenneth I. Shine is president of the Institute of Medicine.

The **National Research Council** was organized by the National Academy of Sciences in 1916 to associate the broad community of science and technology with the Academy's purposes of furthering knowledge and advising the federal government. Functioning in accordance with general policies determined by the Academy, the Council has become the principal operating agency of both the National Academy of Sciences and the National Academy of Engineering in providing services to the government, the public, and the scientific and engineering communities. The Council is administered jointly by both Academies and the Institute of Medicine. Dr. Bruce M. Alberts and Dr. William A. Wulf are chairman and vice chairman, respectively, of the National Research Council.

COMMITTEE ON REMEDIATION OF BURIED AND TANK WASTES

THOMAS M. LESCHINE, *Chair*, University of Washington, Seattle
MARY R. ENGLISH, *Vice Chair*, University of Tennessee, Knoxville
DENISE BIERLEY, Environmental Consultant, St. Helens, Oregon
GREGORY R. CHOPPIN, Florida State University, Tallahassee
JAMES H. CLARKE, Vanderbilt University, Nashville, Tennessee
ALLEN G. CROFF, Oak Ridge National Laboratory, Tennessee
WILLIAM R. FREUDENBURG, University of Wisconsin, Madison
DONALD R. GIBSON, JR., TRW, Colorado Springs, Colorado
NAOMI H. HARLEY, New York University School of Medicine, New York
JAMES H. JOHNSON, JR., Howard University, Washington, D.C.
SHLOMO P. NEUMAN, University of Arizona, Tucson
W. HUGH O'RIORDAN, Givens Pursley, LLP, Boise, Idaho
EDWIN W. ROEDDER, Harvard University, Cambridge, Massachusetts
BENJAMIN ROSS, Disposal Safety Incorporated, Washington, D.C.
RAYMOND G. WYMER, Oak Ridge National Laboratory (retired), Tennessee

Consultants

ROBERT M. BERNERO, U.S. Nuclear Regulatory Commission (retired), Bethesda, Maryland
ELIZABETH K. HOCKING, Argonne National Laboratory, Washington, D.C.
ANNE BALLOU JENNINGS, University of Washington, Seattle

Staff

ROBERT S. ANDREWS, Senior Staff Officer
LAURA D. LLANOS, Senior Project Assistant

BOARD ON RADIOACTIVE WASTE MANAGEMENT

JOHN F. AHEARNE, *Chair*, Sigma Xi and Duke University, Research Triangle Park, North Carolina
CHARLES MCCOMBIE, *Vice-Chair*, Consultant, Gipf-Oberfrick, Switzerland
ROBERT M. BERNERO, U.S. Nuclear Regulatory Commission (retired), Bethesda, Maryland
ROBERT J. BUDNITZ, Future Resources Associates, Inc., Berkeley, California
GREGORY R. CHOPPIN, Florida State University, Tallahassee
JAMES H. JOHNSON, JR., Howard University, Washington, D.C.
ROGER E. KASPERSON, Clark University, Worcester, Massachusetts
JAMES O. LECKIE, Stanford University, California
JANE C.S. LONG, Mackay School of Mines, University of Nevada, Reno
ALEXANDER MACLACHLAN, E.I. du Pont de Nemours & Co. (retired), Wilmington, Delaware
WILLIAM A. MILLS, Oak Ridge Associated Universities (retired), Olney, Maryland
MARTIN J. STEINDLER, Argonne National Laboratory (retired), Argonne, Illinois
ATSUYUKI SUZUKI, University of Tokyo, Japan
JOHN J. TAYLOR, Electric Power Research Institute (retired), Palo Alto, California
VICTORIA J. TSCHINKEL, Landers and Parsons, Tallahassee, Florida
MARY LOU ZOBACK, U.S. Geological Survey, Menlo Park, California

Staff

KEVIN D. CROWLEY, Director
ROBERT S. ANDREWS, Senior Staff Officer
GREGORY H. SYMMES, Senior Staff Officer
JOHN R. WILEY, Senior Staff Officer
BARBARA PASTINA, Staff Officer
SUSAN B. MOCKLER, Research Associate
TONI GREENLEAF, Administrative Associate
LATRICIA C. BAILEY, Senior Project Assistant
LAURA D. LLANOS, Senior Project Assistant
ANGELA R. TAYLOR, Senior Project Assistant
SUZANNE STACKHOUSE, Project Assistant

COMMISSION ON GEOSCIENCES, ENVIRONMENT, AND RESOURCES

GEORGE M. HORNBERGER, *Chair*, University of Virginia, Charlottesville
RICHARD A. CONWAY, Union Carbide Corporation (retired), S. Charleston, West Virginia
LYNN GOLDMAN, Johns Hopkins School of Hygiene and Public Health, Baltimore, Maryland
THOMAS E. GRAEDEL, Yale University, New Haven, Connecticut
THOMAS J. GRAFF, Environmental Defense, Oakland, California
EUGENIA KALNAY, University of Maryland, College Park
DEBRA KNOPMAN, Progressive Policy Institute, Washington, D.C.
BRAD MOONEY, J. Brad Mooney Associates, Ltd., Arlington, Virginia
HUGH C. MORRIS, El Dorado Gold Corporation, Vancouver, British Columbia
H. RONALD PULLIAM, University of Georgia, Athens
MILTON RUSSELL, Joint Institute for Energy and Environment and University of Tennessee (Emeritus), Knoxville
ROBERT J. SERAFIN, National Center for Atmospheric Research, Boulder, Colorado
ANDREW R. SOLOW, Woods Hole Oceanographic Institution, Massachusetts
E-AN ZEN, University of Maryland, College Park

Staff

ROBERT M. HAMILTON, Executive Director
GREGORY H. SYMMES, Associate Executive Director
JEANETTE SPOON, Administrative and Financial Officer
SANDI FITZPATRICK, Administrative Associate

Acknowledgments

This report has been reviewed in draft form by individuals chosen for their diverse perspectives and technical expertise, in accordance with procedures approved by the National Research Council (NRC) Report Review Committee. The purpose of this independent review is to provide candid and critical comments that will assist the institution in making the published report as sound as possible and to ensure that the report meets institutional standards for objectivity, evidence, and responsiveness to the study charge. The review comments and draft manuscript remain confidential to protect the integrity of the deliberative process. We wish to thank the following individuals for their participation in the review of this report:

John S. Applegate, Indiana University
Kenneth Cooke, Bechtel, Inc.
Thomas A. Cotton, JK Research Associates, Inc.
Roy E. Gephart, Pacific Northwest National Laboratory
Roger E. Kasperson, Clark University
Mike Mobley, Tennessee Office of Radiological Health (retired)
Frank L. Parker, Vanderbilt University
Roger W. Staehle, University of Minnesota
Victoria J. Tschinkel, Landers and Parsons

Although the individuals listed above have provided constructive comments and suggestions, they were not asked to endorse the conclusions or recommendations, nor did they see the final draft of the report before its release. The review of this report was overseen by Milton Russell, appointed by the NRC Commission on Geosciences, Environment, and Resources, and by Arden L. Bement, Jr., appointed by the NRC Report Review Committee, both of which were responsible for making certain that an independent examination of this report was carried out in accordance with NRC procedures and that all review comments were carefully considered. Responsibility for the final content of this report rests entirely with the authoring committee and the NRC.

Preface

The Committee on Remediation of Buried and Tank Wastes was asked by the U.S. Department of Energy (DOE) to:

> . . . assess approaches for developing criteria for transition from active to passive remediation and subsequent long-term disposition, including institutional control with monitoring and surveillance, of DOE waste sites and facilities such as Hanford, Washington; Savannah River, South Carolina; Idaho National Engineering Laboratory; and Oak Ridge National Laboratory, Tennessee. Such criteria will include technical feasibility, future land use, performance assessment of remediation activities, and risks to health, safety, and the environment associated with long-term site disposition. Relevant federal and state regulatory requirements and agreements will be included. Appropriate approaches will be applicable to facilities such as high-level radioactive waste tanks (including related facilities and contaminated environments), buried radioactive waste (such as the Hanford low-level waste disposal sites), and on environments contaminated by nuclear testing (such as the Nevada Test Site weapons test event location).

Implicit in this charge is DOE's recognition that radiological and chemical risks are likely to persist at many DOE waste sites for very long time periods, and that protecting humans and the environment from these risks is a dauntingly complex task. For society, now and in the future, this task challenges not only our scientific and technological capabilities, but also our ability to establish and maintain the institutional arrangements that are fundamental to ensuring this protection.

The committee approached its charge by developing a conceptual framework for long-term institutional management of DOE's waste sites. In its study, it concentrated on the sites identified in the DOE request but took other DOE waste sites into account as well. The conceptual framework developed by the committee focuses on three complementary elements of waste site disposition—waste reduction, waste isolation, and stewardship—using the metaphor of a "three-legged stool." The characteristics of and interrelationships among these three elements were examined in the committee's study, as were current capabilities, limitations, and other contextual factors that must be taken into account. Following this assessment, general design criteria for long-term institutional management were identified.

The committee took this general, conceptual approach because the diversity of DOE's waste sites and their residual contaminants, together with large uncertainties about the present and future capabilities of science, technology, and stewardship measures as well as budgetary uncertainties, preclude quantifying the current and future risks posed by various sites or providing a single "recipe for success." Instead, as described in this report, long-term institutional management, broadly and systematically conceived, is essential to responsible site disposition.

In summary, at most of DOE's waste sites complete elimination of unacceptable risks to humans and the environment will not be achieved, now or in the foreseeable future. At many of DOE's sites, radiological and chemical contaminants posing potentially substantial risks are likely to remain on site and may migrate off site. Engineered measures for waste isolation, together with institutional controls and other stewardship measures, will largely be relied upon to prevent unacceptable exposure to these contaminants. The quality of management of residually contaminated waste sites, both in the present and over the longer term, will determine whether these measures are adequately protective. At most sites, no single element—waste reduction, waste isolation, or stewardship—can be relied upon. Long-term institutional management will require an integrated, systems approach that is tailored to the conditions of the site and is revisited over time, as the conditions of the site and its surrounding area change and as new technologies become available.

In closing, we should note the genesis and evolution of this committee. The Committee on Remediation of Buried and Tank Wastes was formed in 1992. Its work has resulted in numerous reports addressing problems of site remediation in the DOE complex (the Idaho National Engineering and Environmental Laboratory aquifer pumping and infiltration test, use of systems analysis and systems engineering at the Hanford Site in Washington, isolation barriers, the Niagara Falls Storage Site, technical management at DOE, and tank waste remediation at Hanford), culminating in the present examination of the long-term disposition of DOE waste sites. This report is in many ways a direct descendant of those earlier studies, and the present committee members owe a debt of gratitude to those earlier members who, though not part of the group that prepared this report, were instrumental in helping to shape the thinking that we brought to bear.

We are indebted to Tom Burke, Bob Catlin, Tom Cotton, Rod Ewing, Glenn Paulson, and, especially, to Paul Witherspoon, whose wit and wisdom have continued to echo through the committee's deliberations. We owe a special debt of thanks to Bob Budnitz, the committee's first chair, whose ceaseless admonitions to "think outside the box" we hope we have honored. The committee also extends its warmest thanks to John Lehr, our DOE liaison, who proved to be a man of infinite forbearance, and to DOE's consultant to the committee, Julie D'Ambrosia, whose insightfulness and ability to sweep aside confusion continue to amaze. All of these people have been of tremendous assistance, but none of them bear responsibility for this report. Finally, we want to acknowledge the essential role that the NRC staff plays in bringing reports like this one to completion. Although they are "just doing their jobs," as Senior Staff Officer Bob Andrews constantly reminded us, Bob and Senior Project Assistant Laura Llanos are especially to be commended for their help and encouragement. We are especially indebted to Bob, who, over the years of the committee's life, became a true friend to all of us.

Thomas M. Leschine, *Chair*
Mary R. English, *Vice Chair*
Committee on Remediation of Buried and Tank Wastes

Contents

SYNOPSIS   1

SUMMARY   3

1 INTRODUCTION   10
   LONG-TERM STEWARDSHIP, 11
   TRANSITION "FROM CLEANUP TO STEWARDSHIP," 12
   PURPOSE OF THE STUDY, 15
   LONG-TERM INSTITUTIONAL MANAGEMENT, 16

2 CONCEPTUAL FRAMEWORK   18
   GENERAL REQUIREMENTS, 18
   SITE DISPOSITION DECISIONS FROM A LONG-TERM INSTITUTIONAL
      MANAGEMENT PERSPECTIVE, 21

3 CONTAMINANT REDUCTION   25
   FUTURE STATES, 27
   CONSTRAINTS AND LIMITATIONS, 32
   FUTURE DIRECTIONS FOR IMPROVEMENTS, 33

4 CONTAMINANT ISOLATION   35
   DESCRIPTION OF THE TECHNOLOGIES, 35
   PERFORMANCE MONITORING OF ENGINEERED BARRIERS AND
      STABILIZED WASTES, 39
   CHARACTERISTICS OF IDEAL CONTAMINANT ISOLATION MEASURES, 40
   CONSTRAINTS AND LIMITATIONS, 41
   FUTURE DIRECTIONS FOR IMPROVEMENT, 41

| 5 | STEWARDSHIP ACTIVITIES | 46 |

COMPONENTS OF A COMPREHENSIVE STEWARDSHIP PROGRAM, 47
TYPICAL INSTITUTIONAL CONTROLS, 50
CONSTRAINTS AND LIMITATIONS, 52
CHARACTERISTICS OF AN EFFECTIVE STEWARDSHIP PROGRAM, 60
FUTURE DIRECTIONS FOR IMPROVING STEWARDSHIP, 61
RELEVANT RESEARCH AND DEVELOPMENT NEEDS, 65

| 6 | CONTEXTUAL FACTORS | 66 |

RISK, 66
SCIENTIFIC AND TECHNICAL CAPABILITY, 68
INSTITUTIONAL CAPABILITY, 69
COST, 70
LAWS AND REGULATIONS, 72
VALUES OF INTERESTED AND AFFECTED PARTIES, 73
OTHER SITES, 74
INTERACTION AMONG CONTEXTUAL FACTORS WITHIN A CLIMATE OF
    UNCERTAINTY, 76

| 7 | FUNDAMENTAL LIMITS ON TECHNICAL AND INSTITUTIONAL CAPABILITIES | 77 |

TECHNICAL CAPABILITIES AND LIMITATIONS, 77
INSTITUTIONAL CAPABILITIES AND LIMITATIONS, 83
BROAD SOCIETAL FACTORS, 86
STRENGTHENING LINKS BETWEEN TECHNICAL AND INSTITUTIONAL
    CAPABILITIES, 89

| 8 | DESIGN PRINCIPLES AND CRITERIA FOR AN EFFECTIVE LONG-TERM INSTITUTIONAL MANAGEMENT SYSTEM: FINDINGS AND RECOMMENDATIONS | 93 |

DESIGN PRINCIPLES AND CRITERIA, 93
FINDINGS, 96
RECOMMENDATIONS, 98

REFERENCES CITED   101

APPENDIXES
  A   COMMITTEE'S STATEMENT OF TASK, 109
  B   CLOSURE PLANS FOR MAJOR DOE SITES, 110
  C   COMMITTEE INFORMATION GATHERING MEETINGS, 120
  D   SUMMARY OF RECENT STEWARDSHIP STUDIES, 125
  E   EXISTING LEGAL STRUCTURE FOR CLOSURE OF THE WEAPONS COMPLEX SITES, 133
  F   DISPOSITION OF THE NEVADA TEST SITE, 141
  G   MATHEMATICAL MODELS USED FOR SITE CLOSURE DECISIONS, 149
  H   BIOGRAPHICAL SKETCHES OF COMMITTEE MEMBERS AND CONSULTANTS, 159
  I   DEFINITIONS OF TERMS USED IN THIS REPORT, 163

FIGURES
1 Map of DOE Nuclear Weapons Complex Sites, 12
2 Long-Term Institutional Management Conceptual Framework, 20

TABLES
1 Institutional Management Characteristics, Criteria, and Principles Found in This Report, 17
2 Summary of Solid Waste Across the DOE Complex, 30

SIDEBARS
1-1 Development of DOE Long-Term Stewardship Report, 13
2-1 Hanford Site Reactor 'Interim Safe Storage,' 23
4-1 Hanford Site Groundwater/Vadose Zone Integration Project, 37
4-2 How Can Radiation Exposures from Waste Disposal be ALARA?, 42
4-3 The Hanford Barrier, 44
5-1 Love Canal, New York: An Example of Failed Stewardship, 53
5-2 The Bikini Atoll Experience: Inherent Fallibility of Institutional Controls and the Virtues of "Defense in Depth," 54
5-3 Institutional Controls at Yucca Mountain Geological Repository, 56
5-4 Trust Funds and Institutional Management, 63
7-1 Role of Models, Site Data, and Science and Technology in Risk Assessment and Management, 80
7-2 Evaluation of Nevada Test Site Groundwater Modeling, 81
7-3 Reindustrialization of the Mound Site, 87
7-4 Basic Research Needs in Subsurface Science, 90

Synopsis

This study examines concerns raised by the U.S. Department of Energy (DOE) in its planning for transition from active waste site management and remediation to what the department terms "long-term stewardship." It examines the scientific, technical, and organizational capabilities and limitations that must be taken into account in planning for the long-term institutional management of the department's numerous waste sites that are the legacy to this country's nuclear weapons program. It also identifies characteristics and design criteria for effective long-term institutional management.

Of the sites in DOE's inventory, few will be cleaned up sufficiently to allow unrestricted use. At many sites, radiological and non-radiological hazardous wastes will remain, posing risk to humans and the environment for tens or even hundreds of thousands of years. In some cases, contaminants have migrated off site or are likely to do so in the future. Future changes in the uses of sites and nearby areas make predicting risks even more difficult. In response to the technological, budgetary, and societal problems posed by these sites, DOE plans to rely on institutional controls and other stewardship measures to prevent exposure to residual contaminants following activities aimed at stabilization and containment. One message that emerges from this study, however, is that effective long-term stewardship will likely be difficult to achieve.

In this study it is argued that, while stewardship as defined by DOE is essential, a much broader-based, more systematic approach is needed. For any given site, contaminant reduction, contaminant isolation, and stewardship should be treated as an integrated, complementary system: one that requires foresight, transparently clear and realistic thinking, and accountability. Today's waste management actions should become an integral part of stewardship planning. Scientific, technical, and organizational deficiencies or knowledge gaps should be acknowledged frankly and, where possible, research investments should be made to correct them. The long-term institutional management plan for a legacy waste site should strive for stability, balanced by flexibility and provisions for iteration over time. No plan developed today is likely to remain protective for the duration of the hazards. Instead, long-term institutional management requires periodic, comprehensive reevaluation of those legacy waste sites still presenting risk to the public and the environment to ensure that they do not fall into neglect and that advantage is taken of new opportunities for their further remediation.

Summary

It is now becoming clear that relatively few U.S. Department of Energy (DOE) waste sites will be cleaned up to the point where they can be released for unrestricted use. "Long-term stewardship" (activities to protect human health and the environment from hazards that may remain at its sites after cessation of remediation) will be required for over 100 of the 144 waste sites under DOE control (U.S. Department of Energy, 1999). After stabilizing wastes that remain on site and containing them as well as is feasible, DOE intends to rely on stewardship for as long as hazards persist—in many cases, indefinitely. Physical containment barriers, the management systems upon which their long-term reliability depends, and institutional controls intended to prevent exposure of people and the environment to the remaining site hazards, will have to be maintained at some DOE sites for an indefinite period of time.

The Committee on Remediation of Buried and Tank Wastes finds that much regarding DOE's intended reliance on long-term stewardship is at this point problematic. The details of long-term stewardship planning are yet to be specified, the adequacy of funding is not assured, and there is no convincing evidence that institutional controls and other stewardship measures are reliable over the long term. Scientific understanding of the factors that govern the long-term behavior of residual contaminants in the environment is not adequate. Yet, the likelihood that institutional management measures will fail at some point is relatively high, underscoring the need to assure that decisions made in the near term are based on the best available science. Improving institutional capabilities can be expected to be every bit as difficult as improving scientific and technical ones, but without improved understanding of why and how institutions succeed and fail, the follow-through necessary to assure that long-term stewardship remains effective cannot reliably be counted on to occur.

Other things being equal, contaminant reduction is preferred to contaminant isolation and the imposition of stewardship measures whose risk of failure is high. While DOE can do much to assure that stewardship considerations become more pervasive in all aspects of DOE operations, many of the limitations in current capabilities pointed to in this report will likely require higher-level attention. Prominent among these are assured funding for long-term institutional management. Moreover, the current regulatory framework for waste site remediation appears to encourage a constrained and piecemeal approach that makes it difficult to assure that the broader needs of effective long-term institutional management get the consideration they deserve.

This study examines the capabilities and limitations of the scientific, technical, and human and institutional systems that compose the measures that DOE expects to put into place at potentially hazardous, residually contaminated sites. The committee finds that, at a minimum, DOE should plan for site disposition and stewardship much more systematically than it has to date. At many sites, future risks from residual wastes cannot be predicted with any confidence, because numerous underlying factors that influence the character, extent, and severity of long-term risks are not well understood. Among these factors are the long-term behavior of wastes in the environment, the long-term performance of engineered systems designed to contain wastes, the reliability of institutional controls and other stewardship measures, and the distribution and resource needs of future human populations.

Because uncertainty is inherent in many of these areas, and because DOE's preferred solutions—reliance on engineered barriers and institutional controls—are inherently failure prone, step-wise planning for DOE legacy sites must be *systematic, integrative, comprehensive, and iterative* in its execution through time, *adaptive* in the face of uncertainty, and *active* in the search for new and different solutions. Planning for long-term institutional management should commence while remediation is underway. Ideally, its needs are taken into account as facilities are being designed and waste management operations initiated.

To the extent that long-term stewardship imposes costs and risks on future generations, questions of intergenerational equity are raised that should be recognized in current planning. Waste site remediation is appropriately left to future generations if risks are low, if it is impractical with currently available technology, or if it would impose unacceptable costs on society were it to be undertaken today. Remediation is inappropriately left to future generations if the risks are such that what is a tractable remediation problem today becomes much less so in the future as a result of events or changes in conditions that could reasonably have been foreseen. Unfortunately, for most waste sites, little information is presently available that facilitates well-considered examination of such tradeoffs. To the extent that long-term institutional management becomes a logical extension of today's waste management activities, as the committee believes it should, the need to confront such difficult tradeoffs should lessen. Developing new facilities and managing today's wastes with the needs of long-term stewardship in mind is an important aspect of the integrative approach embodied in the committee's framework for long-term institutional management.

This study uses the term *long-term institutional management* to refer to a planning and decision-making approach that strives to achieve an appropriate balance in the way it employs contaminant reduction measures, engineered barriers that isolate residual contaminants from the human environment and retard their migration, and places reliance on institutional controls and other stewardship measures. Decisions are guided by consideration of contextual factors that include:

- risks to members of the public, workers, and the environment;
- legal and regulatory requirements;
- technical and institutional capabilities and limitations, and the current state of scientific knowledge;
- values and preferences of interested and affected parties;
- costs and related budgetary considerations; and
- impacts on and activities at other sites.

To the extent that the above contextual factors constrain decisions, a well-functioning long-term institutional management system works to curtail those constraints that compromise the basic goal of containing and minimizing the risks that prevent unrestricted release of DOE sites.

The limitations of "hardware" systems and supporting scientific understanding are amplified by the inherent fallibility of the human and organizational systems upon which stewardship ultimately depends. For this reason, emphasis is placed in this report on the management systems for long-term planning and decision making at individual DOE sites. The report recommends that DOE apply five planning principles to the management of residually contaminated sites: (1) plan for uncertainty, (2) plan for fallibility, (3) develop appropriate incentive structures, (4) undertake necessary scientific, technical, and social research and development, and (5) plan to maximize follow-through on phased, iterative, and adaptive long-term institutional management approaches. For this purpose, a long-term commitment to both basic and applied research is needed. This research must address not only improvement of technical and human systems performance, but also basic scientific questions about the behavior of wastes in the diverse environments of the nation's nuclear waste sites. While there is no assurance that management systems will continue to be effective for the future, even short-term effectiveness cannot be assured without continued, adequate funding.

Numerous measures are necessary to assure that the integrity of engineered barriers intended to isolate wastes from the environment is maintained, that the behavior of unconfined wastes in the environment is as expected, and that unanticipated exposure pathways to humans or other sensitive species do not develop. Experience to date, both at DOE sites and at hazardous waste sites elsewhere, suggests that the tools available for these purposes are of doubtful technical effectiveness. The building of an effective long-term program for DOE legacy waste sites poses a substantial challenge to "remediation technology," broadly construed. It challenges the basic science upon which technological advance depends, as well as the knowledge of organizational and human behavior upon which our ability to design effective long-term management systems ultimately rests.

The committee believes that the working assumption of DOE planners must be that many contamination isolation barriers and stewardship measures at sites where wastes are left in place will eventually fail, and that much of our current knowledge of the long-term behavior of wastes in environmental media may eventually be proven wrong. Planning and implementation at these sites must proceed in ways that are cognizant of this potential fallibility and uncertainty.

How site planning and management should proceed, given this working assumption, is a primary focus of this report. DOE has not as yet developed in any detail the institutional arrangements through which long-term site management would be implemented. Nor have these arrangements been discussed very much among DOE and its partners in state and federal regulatory agencies, site host communities, affected Indian tribes, and environmental organizations. It is important that DOE involve its Site Specific Advisory Boards in its long-term stewardship planning as early as possible. Although the rationale for long-term stewardship at DOE waste sites has been put forward in a general way in several recent studies (Probst and McGovern, 1998; U.S. Department of Energy, 1999), no coherent framework for long-term planning at individual DOE waste sites has as yet emerged. This report tackles the question of the character of the management systems that the committee believes are necessary, applying information gleaned from numerous sites to develop a general conceptual approach that can be applied on a site-specific basis. While complex-wide integration and planning are also needed, the committee's framework is intended to apply primarily on the individual, site-specific level.

WHAT IS LONG-TERM INSTITUTIONAL MANAGEMENT OF WASTE SITES?

Long-term institutional management is the committee's conception of an approach to planning and decision making for the management of contaminated sites, facilities, and materials. It represents the framework in which tradeoffs among contaminant reduction, reliance on contaminant isolation, and stewardship measures are made. The framework represents a synthesis of the committee's examination of what is and is not likely to work in long-term waste site management. It incorporates the measures available to site managers as remediation or stewardship planning moves forward, the factors that influence the site management choices made at particular points in time, and the iterative character of decision making through time as new information emerges or planned site end state goals are adjusted.

The committee's metaphor for balancing the three basic elements that waste-site managers have at their disposal—contaminant reduction, physical isolation of residual contaminants, and deployment of stewardship activities—is a "three-legged stool." These three basic sets of measures are represented by the stool's "legs." The goals or end state they are trying to achieve are represented by the stool's "seat," and the contextual factors listed earlier that constrain their use are represented by the "rungs." Metaphorically, the rugged terrain upon which the stool rests represents the variability of contamination scenarios within and among sites. This framework is developed in anticipation of the numerous questions DOE will face as it develops long-term plans for contaminated sites. *In all cases reviewed by the committee, current DOE remediation planning and planning for post-remediation stewardship can fit within the conceptual framework developed in this study. In no case, however, was planning and management as highly developed as the committee's framework suggests it should be.*

WHY IS LONG-TERM INSTITUTIONAL MANAGEMENT NECESSARY AT DOE WASTE SITES?

For reasons that are technical, social, fiscal, and political, most DOE sites will not be cleaned up well enough to allow unrestricted release of the land. In a few cases the rationale for leaving contaminants in place includes a judgment that the collateral environmental damage of available remediation technologies outweighs the benefits likely to be achieved. According to recent departmental estimates, 109 of the 144 DOE waste sites, including its largest sites (such as the Hanford Site in Washington, Oak Ridge Reservation in Tennessee, Savannah River Site in South Carolina, and Idaho National Engineering and Environmental Laboratory) are unlikely to become available for site-wide unrestricted use (U.S. Department of Energy, 1999). *The large inventory of sites requiring long-term management, the nature and complexity of many of these sites, coupled with the limitations of subsurface science, requires comprehensive and systematic planning that embraces the principles of long-term institutional management described in this report.*

The fiscal limitations that preclude more complete remediation are largely a matter of national policy. At some sites the preferred land uses following completion of DOE's mission are still being debated, while at others the future roles of the sites are under discussion (Probst and Lowe, 2000). Total cleanup costs are very sensitive to the nature of the cleanup end states selected, with large increments in estimated costs associated with moving sites from a restricted-access "iron fence" condition to the point where they can be released for unrestricted use (U.S. Department of Energy, 1996). Roughly $50 billion has been spent on remediation to date; a recent report prepared by the U.S. Department of Energy (2000b) estimates that the life-cycle costs yet to be incurred are approximately $151 to $195 billion.

By contrast, DOE officials view the long-term stewardship efforts, which are likely to rely heavily on land control, site surveillance, monitoring, maintenance, record keeping, and related activities, as inherently low cost. *The real long-term costs of site stewardship cannot be estimated with any confidence, however. Even after the details of a comprehensive long-term institutional management plan are in place, large uncertainties are likely to cloud true economic costs. In addition, equating long-term management costs with the costs of the specific stewardship activities envisioned over as long a peroid as several thousands of years fails to account for the societal costs of stewardship system failures (e.g., aquifers becoming contaminated by residual wastes whose propensity for off-site migration was not understood at the time active remediation ended). A well-designed long-term institutional management system should have as a goal the anticipation of stewardship failures and minimization of the costs and risks associated with them. It accomplishes this through investment in improving the management system itself, and in improved scientific understanding and improved remediation technology, each of which is capable of reducing these potentially large costs and risks to society in the future.*

At the larger DOE sites where local economic, political, and environmental factors already exert a strong influence on site decision making, the necessity for an integrated and forward-looking approach to long-term planning becomes especially clear. For example, growth in the Denver metropolitan region that is encroaching upon the Rocky Flats site, or the rapidly growing Las Vegas area that might one day look to areas around the Nevada Test Site for water. A different approach to long-term institutional management planning might be appropriate for sites where significant changes in the pattern of future uses are less likely. However, projections of future land uses and the values of members of the public must receive careful consideration, no matter where the site is located. At some sites, subsurface contaminants are now known to be migrating further from their sources than originally predicted, with future consequences that are not well understood at present.

IMPLICATIONS OF SCIENTIFIC, TECHNICAL, AND INSTITUTIONAL CAPABILITIES AND LIMITATIONS FOR LONG-TERM INSTITUTIONAL MANAGEMENT

The site management measures that DOE has at its disposal, whether they are the "hardware" systems used for waste remediation and containment or the institutional systems under which all site activities occur, share the characteristic of being limited in what they can accomplish. Were contaminant reduction efforts able to perform at anything like their theoretical ideal, many of the site custodianship problems that DOE now faces would disappear. As a general rule, however, the greater the degree of decontamination, the greater the cost and, in some cases, the greater the worker risk and adverse environmental effects. Groundwater contamination is pervasive at DOE sites, and "pump and treat" operations, whether intended to reduce contamination levels or to retard migration, are expected to run for decades—or even centuries—to achieve their desired results.

In some cases, the lack of sufficient pre- or post-remediation characterization of either the wastes or the environments into which they have been placed can render realistic estimation of the effectiveness of contaminant reduction measures nearly impossible. A key question for each site must be "How much characterization is sufficient to overcome this impasse?" A major concern is the adequacy of understanding of the physical and chemical properties of the environment in which contaminants reside and their transport through the environment over time. Mathematical modeling of contaminant fate and transport is an essential tool for long-term institutional management, but its track record to date at DOE sites, particularly where contaminants reside in the unsaturated, or "vadose" zone, has been mixed. This necessitates integration of a science and technology program into both site remediation planning (National Research Council, 2000b) and the activities that follow after remediation activities cease.

In situ engineered barriers are likely to be widely applied as the need for them is closely coupled to the extent to which contaminant reduction measures are effective. Once in place, the ongoing effectiveness of the systems that are emplaced to isolate and prevent the movement of contaminants depends on institutional management, typically in the form of monitoring and maintenance. Knowledge of the effective lifetimes of the materials and systems used in barrier design is limited, however, and comparatively little performance monitoring data exists. *The lack of experience with the long-term performance of engineered barriers, coupled with the heavy reliance being placed upon them at DOE sites, is another factor that necessitates an approach to long-term institutional management that actively seeks out and applies new knowledge.*

In situ barriers used to isolate long-lived contaminants from the environment will have to be not only maintained, but in some instances completely replaced. Initial emplacement of barrier systems must therefore take that possibility into account. *Irrespective of the management systems put in place in support of other aspects of long-term stewardship programs, physical barrier systems to keep hazardous wastes in isolation will require their own ongoing support from the institutional management system.*

Stewardship in its broadest sense includes all of the activities that will be required concerning potentially harmful contamination left on site following the completion of remediation. The issues for long-term institutional management include not only what will be done, but how, and when, and by whom. Institutional controls, often especially important elements of stewardship, consist mainly of land use or access restrictions, and they can take the form either of legal restrictions imposed through covenants, easements, and the like, or of physical restrictions, such as fences, warning signs, or the posting of guards. Stewardship is not limited to institutional controls, however. It also includes information management and dissemination, oversight and enforcement, monitoring and maintenance, periodic reevaluation of protective systems, and cultivating new remediation options.

Without constant attention, stewardship measures imposed today are not likely to remain effective for as long as residual contamination presents risks. It will, however, be very difficult to assure that proper attention continues over time. This means that stewardship and science—both basic science and applied science and technology research and development—are interdependent and must be managed together. *Site stewardship that includes the monitoring and encouragement of emerging new technologies and scientific breakthroughs for their relevance to further reducing the risks associated with residual contaminants would, over the long run, decrease the potential consequences of stewardship failures.*

Many weaknesses in institutional controls and other stewardship activities stem from inherent institutional fallibilities. Understanding and predicting the nature and pervasiveness of institutional fallibility, particularly where long-term attention to mission is required, is essential if the organizations charged with long-term management of waste sites are to be designed in ways that make them resistant to failures that compromise the safety of sites with residual wastes. *Because the organizational systems charged with long-term care and custodianship of hazardous materials and for some types of public goods have proven so fallible in the past, the research and development efforts that are part of long-term institutional management need to extend to the social, institutional, and organizational aspects of long-term management systems as well.*

"BIGGER PICTURE" FACTORS THAT ARGUE FOR A LONG-TERM INSTITUTIONAL MANAGEMENT APPROACH

Long-term institutional management decisions are often constrained by contextual factors not easily controllable by site managers. These include risks, the state of scientific understanding, technical and institutional capabilities, costs, laws and regulations, the views of interested and affected parties, and activities at other sites. The latter includes nearby contaminated sites, nearby lands outside the facility, receptor sites, and similar sites, particularly similar sites within the DOE complex.

The status of lands around a contaminated site, including the presence of other contaminated sites nearby, can strongly affect site disposition decisions. Often, however, the separation of sites for administrative purposes (e.g., into operable units or solid waste management units) conflicts with the logic suggested by a site's natural geography, hydrology, and geology. Changing land uses or resource consumption patterns beyond the administrative boundaries of a site, but within its natural environment, can both affect and be affected by the conditions of the site. Human-induced changes in hydrologic conditions, for example, may affect the ability of isolation technologies to keep soil contaminants out of groundwater. The combination of changing human demand for water, coupled with the induced change in the availability of contaminants to the same groundwater system, can thus create risks that might not otherwise exist. *Successful management of risks will require that the institutional management system be able either to anticipate and prevent such problems before they occur, or to detect and reverse the underlying changes before harm is done.* Whether either of these can be done reliably over the long term is open to question.

One way to attempt to overcome both technical and institutional limitations is to forge links between technical and institutional capabilities. The two can be mutually reinforcing in (1) the periodic reevaluation of site disposition decisions, and (2) the development of new technologies that lessen the dependence on fallible institutional arrangements that were necessitated by the technical limitations of the past.

DESIGNING AND IMPLEMENTING A SITE'S INSTITUTIONAL MANAGEMENT SYSTEM

General design criteria exist that can help assure that a site's system of institutional management reflects an appropriate balance in the reliance it places on each of the three "legs" of the long-term institutional management "stool." Nine such criteria (discussed in Chapter 8) emerge from this study.

• *Defense in depth* refers to layering by using more than one measure to accomplish basically the same purpose, and redundancy by having more than one organization responsible for basically the same task.

• *Complementarity* refers to the support that each measure provides to the others.

• *Foresight* refers to the ability, despite uncertainties, to anticipate how the components of the system will or will not work individually and as a whole. Adjustments are then made beforehand or contingencies planned for accordingly.

• *Accountability*, which extends to both the public and government authorities, requires both a willingness to be made answerable and the technical means to identify and correct performance defects.

• *Transparency* means that the basis for site management decisions is clear and that the public has the opportunity to review and comment on these decisions before they are finalized. Transparency lays the groundwork for accountability.

• *Feasibility* refers to having an institutional management system that is technically, economically, and institutionally possible to implement within a specific time period.

• *Stability through time* refers to the likelihood that, based on reasonable estimates, the individual components of the site management system and the system as a whole will continue to perform as initially configured.

• *Iteration* refers to the concept that the whole system requires periodic reexamination to determine whether the various parts of a site's protective system are functioning as expected and whether system performance can be improved.

• *Follow-through* and *flexibility* refer to a commitment to taking innovative action to correct or redirect a site's management system when a need is identified.

In addition to these design criteria, there are other characteristics that institutional management systems should have that fall into the category of implementation criteria—that is, attributes of the system that, if included, increase chances that it will be successfully implemented and maintained over time. These include:

- *Clear objectives* and a desire on the part of those responsible for institutional management to carry out those objectives with *diligence over time.*
- A *clear system of governance* that specifies what is to be done and by whom and is founded on precepts that are enduring on the one hand and flexible on the other.
- An *integrated overall approach* that coordinates activities across the responsible entities and assures that site management measures are complementary rather than conflicting.
- *Incentives* both within and outside the institutional management organization to encourage diligence in carrying out mission objectives.

The mechanisms for creating and implementing effective long-term institutional management do not necessarily have to be created "from scratch." Some mechanisms with at least some of the attributes mentioned here already exist, both within and outside of DOE, and others, such as the program within the DOE Environmental Management Office of Long-Term Stewardship, are coming into being. Nevertheless, a systematic approach is needed for the many challenges that such mechanisms will have to face to be overcome. By the same token, a number of other factors that do not appear as specific characteristics of institutional mechanisms are essential to maintain their effectiveness through time. These include, for example, positive incentive structures that encourage system personnel to behave in ways that reinforce the management system's basic purpose, and stable funding through time.

In conclusion, given that unrestricted use will not be possible for many DOE legacy waste sites, and given that decisions that affect sites' futures are often made under conditions of considerable uncertainty, the best decision strategy overall appears to be one that avoids foreclosing future options where sensible, takes contingencies into account wherever possible, and takes seriously the prospects that failures of engineered barriers, institutional controls, and other stewardship measures in the future could have ramifications that a good steward would want to avoid. A forward-looking strategy is essential because today's scientific knowledge and technical and institutional capabilities are insufficient to provide much confidence that sites with residual risks will continue to function as expected for the time periods necessary. "Cookbook" approaches are unlikely to be successful, and there is no "one size fits all" formula for successful institutional management. In designing long-term institutional management systems, flexibility, equity, efficiency, and environmental and human health protection objectives must be attended to, more or less simultaneously. Management strategies that are iterative and provide "follow-through" on these objectives over time enhance the chances that the ultimate health and safety objectives will be met.

1

Introduction

One of the most prescient comments of the nuclear age is Alvin Weinberg's (1972) now classic observation:

> We nuclear people have made a Faustian bargain with society. On the one hand, we offer, in the catalytic nuclear burner, an inexhaustible source of energy. . . . But the price that we demand of society for this magical energy source is both a vigilance and a longevity of our social institutions that we are quite unaccustomed to.

While Weinberg's comment referred to spent nuclear fuel, it is applicable to the products and byproducts of nuclear weapons production as well. These observations take on added importance, however, now that it has become clear that many U.S. Department of Energy (DOE) sites will require ongoing management for very long periods of time. The department has recently estimated that the great majority of sites currently in its care will not be able to be cleaned up to the point where they can be released for unrestricted use (U.S. Department of Energy, 1999). Factors such as technical infeasibility, excessive worker risk or environmental damage, the explicit choices that are made, and costs dictate the extent to which sites are undergoing remediation. Sites expected to require what the DOE has termed "long-term stewardship" are found in more than half of the states, in Puerto Rico, and in the Trust Territory of the Pacific. For many locations within the DOE defense complex,[1] a group that includes the largest and most contaminated sites in this country, the terms of long-term stewardship have yet to be specified in any detail.

Due to the nature of the hazards involved, the stewardship measures at many sites, once instituted, will have to be maintained for long periods of time. For this reason the primary focus of this report is on the need for *long-term institutional management* of contaminated sites—the attributes that long-term management must have to be effective and the conditions necessary for its establishment at these contaminated sites. The committee's conception, detailed in Chapter 2, views the selection, implementation, and periodic reassessment of stewardship mea-

---

[1] The use in this report of the term "DOE defense complex" or "DOE legacy waste sites" refers to those areas making up a contiguous block of land owned or managed by DOE and containing radioactive and hazardous wastes that are the legacy of nuclear weapons production. According to the report by the U.S. Congress Office of Technology Assessment (1991) entitled *Complex Cleanup*, work performed within the complex has included: (1) weapons research and development; (2) nuclear materials (plutonium and tritium) production and processing, along with uranium processing; (3) warhead component production; and (4) warhead testing. DOE program wastes also include nuclear energy, isotope production, and nuclear propulsion.

sures as important functional components of long-term institutional management. They serve as complements and supplements to the contaminant reduction and isolation activities that must continue to be addressed as part of institutional management for as long as unacceptable hazards persist.

LONG-TERM STEWARDSHIP

The term "long-term stewardship" is used by DOE to describe the care and attention that contaminated areas will receive after cleanup is "complete"[2] (remediation ends). The need for attention to post-remediation site controls is perhaps best appreciated by considering the goals for site completion that DOE has set for cleanups across the complex. According to the report *Accelerating Cleanup: Paths to Closure* (U.S. Department of Energy, 1998a, pp. 1-7), DOE considers a site to be "complete" (or at its end state) when:

- Deactivation or decommissioning of all facilities currently in the EM [DOE Office of Environmental Management] program has been completed, excluding any long-term surveillance and monitoring;
- All releases to the environment have been cleaned up in accordance with agreed-upon cleanup standards;
- Groundwater contamination has been contained, or long-term treatment or monitoring is in place;
- Nuclear material and spent fuel have been stabilized and/or placed in safe long-term storage; and
- "Legacy" waste (i.e., waste produced by past nuclear weapons production activities, with the exception of high-level waste) has been disposed of in an approved manner.

Stewardship activities, like those described in the department's recent report, *From Cleanup to Stewardship* (U.S. Department of Energy, 1999), thus become relevant when potentially harmful contaminants remain after remediation has been "completed." The importance of such situations to the nation as a whole is apparent: relatively few DOE sites will, in the foreseeable future, be cleaned up to the point where no post-remediation measures are necessary (i.e., the highly desirable end state condition where no restrictions on future use are needed). The reasons for this are technical, financial, social, and political. DOE estimates that 109 of the 144 sites currently under its care will require some kind of protective stewardship after currently planned remediation activities are complete (U.S. Department of Energy, 1999).[3]

Some sites requiring long-term stewardship are located in close proximity to human populations (e.g., the Mound Plant in Ohio), while others, once fairly isolated, are now being encroached upon by local population growth (e.g., the Rocky Flats Site near Denver, Colorado). Figure 1 shows the locations of the 109 DOE sites mentioned above that will require stewardship, and Appendix B gives a brief summary of closure plans for the major DOE legacy sites as described in various reports by the U.S. Department of Energy (1995a, 1996, 1998a, 1999).

It is difficult to determine from available documentation where stewardship needs are most pressing because the final end states (or conditions) of the sites are not reliably known and the activities that will constitute stewardship have yet to be defined. The report, *Linking Legacies* (U.S. Department of Energy, 1997b), gives some clues. Of the total amount of legacy waste radioactivity (over 1 billion curies), the largest amounts (about 86 percent of the total) are found as high-level waste at DOE sites that performed chemical separations—Savannah River Site in South Carolina, Hanford Site in Washington, and the Idaho National Engineering and Environmental Laboratory. In terms of waste volume (about 36 million cubic meters), about 89 percent is byproduct material (mill

---

[2] In its December 14, 1998, settlement of a lawsuit (Natural Resources Defense Council, et al. v. Richardson, et al., Civ. No. 97-963 [SS]) the department defined long-term stewardship as: "the physical controls, institutions, information and other mechanisms needed to ensure protection of people and the environment at sites where DOE has completed or plans to complete 'cleanup' (e.g., landfill closures, remedial actions, removal actions, and facility stabilization). This concept ... includes, *inter alia*, land use controls, monitoring, maintenance, and information management." (*Federal Register*, October 6, 1999, vol. 64, no. 193, p. 54280).

[3] The term "site," though often used in this report in reference to whole sites, is used by DOE to refer more generally to "geographically distinct locations [within whole sites] as well as specific disposal cells, contained contamination areas, and entombed contaminated facilities" (*Congressional Record*, August 5, 1999, p. H7855). The figure quoted thus refers to the fact that portions of 109 sites in the weapons complex will require protective stewardship despite whole-site cleanup being regarded as "complete."

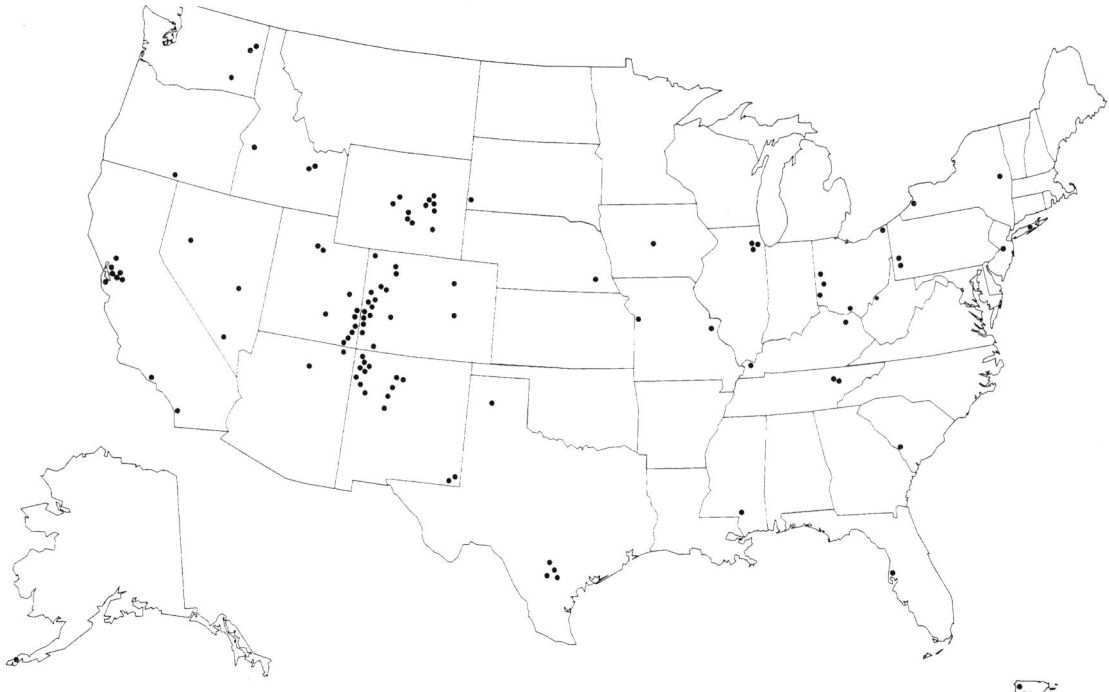

FIGURE 1 Map of DOE nuclear weapons complex sites (from U.S. Department of Energy, 1999).

tailings or waste produced by the extraction of uranium or thorium from source ores) stored at burial sites located primarily in the western U.S.; this waste contains less than 1 percent of the total radioactivity.

TRANSITION "FROM CLEANUP TO STEWARDSHIP"

The department's *Baseline Environmental Management Reports* (U.S. Department of Energy, 1995a, 1996) and *Accelerating Cleanup: Paths to Closure* report (U.S. Department of Energy, 1998a) were intended primarily to aid in the scheduling and budgeting of the cleanup. But they also called attention to the need for protective measures long after planned remediation activities were completed. Various reasons for reliance on long-term protective measures were presented to the committee in site-by-site assessments, but principal themes were the technical infeasibility of remediation (e.g., long-term pump-and-treat efforts at most sites; see Appendix B) or the collateral environmental damage that would be entailed if contaminants were to be physically removed (e.g., attempts to remove radiocesium, mercury, and polychlorinated biphenyls from sediments in the Clinch River, Tennessee, might result in new exposures to the public and the environment). In other cases, consideration of residual risks and the costs of removal dictated that wastes could be safely left in place if isolation caps or other protective barriers were properly constructed and maintained (e.g., this was the decision reached concerning uranium mill tailings left at many former mining and milling operations in the western U.S.).

In its report, *From Cleanup to Stewardship* (U.S. Department of Energy, 1999), the DOE made its first effort to detail how individual sites in the complex will reach a condition in which long-term stewardship is the main activity. This document outlines a long-term stewardship study that DOE agreed to prepare by the end of the year 2000 when it settled a long-standing lawsuit brought by the Natural Resources Defense Council. As described in a recent *Federal Register* notice (see footnote 2), this study will lay out complex-wide long-term stewardship issues and challenges, define policy options, and detail the department's long-term responsibilities. A recent

> **SIDEBAR 1-1**
>
> **DEVELOPMENT OF DOE LONG-TERM STEWARDSHIP REPORT**
>
> On January 24, 2000, the U.S. Department of Energy (DOE) Office of Environmental Management (EM) Headquarters issued a guidance document to its operations and field offices for their preparation and submission of information concerning ongoing and long-term stewardship activities at DOE sites. This information is to be combined into a National Defense Authorization Act Long-Term Stewardship Report requested by the Congress. It was requested for two reasons:
>
> 1) The report responds directly to a congressional mandate pursuant to the 1998 agreement for the lawsuit settlement of National Resources Defense Council et al., versus Richardson et al., Civ. No. 97-963 (SS) (D.D.C. December 12, 1998). Of concern to Congress is that DOE/EM needs to demonstrate what has been accomplished with the nearly $60 billion provided to its program during the last 10 years.
>
> 2) Congressional staff and state and local governments have expressed interest in long-term stewardship planning, responsibility, and activities.
>
> In developing these information submittal protocols, the guidance document strongly recommends that each field office involve the public. The resulting report is for planning purposes only and in no way indicates any preferences or preempts any ongoing or future regulatory process. Assumptions are to be documented and estimates are to be based on the best available understanding of a given site. The information is to include site descriptions, missions, and cleanup goals; details on the contaminated portions of each site (characterization of the contaminants and the surrounding environment); long-term stewardship goals and activities; estimated long-term stewardship costs; and future land uses.
>
> **REFERENCES**
>
> U.S. Department of Energy. 2000a (January 24). Guidance for the Development of the FY 2000 National Defense Authorization Act (NDAA) Long-Term Stewardship Report. Office of Environmental Management, Washington, D.C.

document from the U.S. Department of Energy (2000a) outlines data for such a report requested by DOE Headquarters from its operations and field offices (see Sidebar 1-1).

The fiscal year 2000 National Defense Authorization Act directs DOE to prepare, on a site-specific basis, a report on existing and anticipated long-term stewardship responsibilities, including cost estimates where available. Congress intends that such a study address those individual sites (or portions thereof) for which cleanup will be completed by the end of the year 2006 (*Congressional Record*, August 5, 1999, p. H7855). Due in October 2000, that study may make apparent for the first time the full scope of the department's long-term stewardship obligations. During the course of the committee's study, DOE established a Long-Term Stewardship Information Center that provides information on the long-term care of DOE sites to the interested public.

These recent trends should be viewed in the broader context of the DOE cleanup program. Planning for the cleanup of the numerous sites and facilities that comprise the nation's nuclear weapons complex got underway during the 1980s, following a series of court cases that clarified the responsibilities of DOE and the oversight roles of other agencies and host-state governments (U.S. Congress Office of Technology Assessment, 1991). Management oversight for the cleanup was centralized with the creation of the DOE Office of Environmental Management

(EM) in 1989. In the ten years following, cleanup planning has, in the view of some observers, progressed through two distinct stages (Bjornstad, Jones, and Dümmer, 1997).

In the first stage, the main priority was achieving the greatest possible degree of cleanup, with planning largely driven by the aspirations of managers and operators of the individual sites. The second and current stage commenced in 1996 with the release of what DOE/EM then referred to as the "2006 Plan" (Alm, 1997). Under this plan and subsequent revisions, budgetary pressures from DOE headquarters have gained in influence. With the U.S. Congress and administration both concerned about the growing future costs implied by the DOE *Baseline Environmental Management Reports* (U.S. Department of Energy, 1995a, 1996),[4] cost-savings objectives have become increasingly important. In addition, host communities whose economic well-being was strongly tied to the DOE Cold War-era defense mission have argued that DOE should provide transition assistance, especially as the economic prospects associated with the cleanup also began to dim (Russell, 1997).

Alongside these budgetary constraints, it has been recognized that at least some contamination problems within the complex simply cannot be cleaned up with currently available technology (U.S. Department of Energy, 1996, Table 3.1). All these changes in thinking have led DOE to plan to close many sites with large inventories of contamination left in place (U.S. Department of Energy, 1998a, 1999), and with remediation approaches that may not be durable over the long time period that the contaminant problems will persist. Appendix B of this report summarizes the current planning efforts.

The DOE/EM current cleanup budget of about $6 billion per year has not grown appreciably in several years. DOE's own analysis suggests that the total cost of cleanup is sensitive to the cleanup goals selected for contaminated sites, a point that is illustrated in the 1996 *Baseline Environmental Management Report*. By far the biggest cost increment between scenarios occurs when a "modified greenfields" scenario (in which the most contaminated areas within the five largest sites are left in a condition requiring highly restricted access) is replaced by what is termed a "maximum feasible greenfields" scenario.[5]

DOE's current emphasis is on completing cleanups and closing sites where this can be done relatively soon. The recent DOE/EM reorganization plan creates a new entity, the Office of Long-Term Stewardship, a subdivision of the Office of Science and Technology, to which long-term stewardship is assigned. Its function is to develop policy and research for DOE's stewardship activities, carried out operationally by the DOE Grand Junction Office in Colorado. The reorganization plan seems to signal that the DOE/EM emphasis is shifting toward project completion and site closure. With the cleanup of the largest sites in the complex to the completion condition described in *Accelerating Cleanup: Paths to Closure* (U.S. Department of Energy, 1998a) currently projected to take as long as 70 years, long-term institutional management measures will be needed at many sites even while remediation continues.

For the larger, more complex DOE sites, the costs associated with long-term stewardship have yet to be estimated in any detail. Departmental officials who briefed the committee expressed the view that the annual operational and maintenance costs associated with site stewardship would likely be relatively low in comparison with the current costs of site operations or site cleanup (J. Werner, DOE/EM, personal communication, 1999). However, no reliable cost projections are currently available for the expenditure that many years of stewardship will require. Formal procedures for the transfer of sites from environmental remediation to long-term stewardship have yet to be established.

---

[4] The base-case life-cycle cost estimate from 1997 through 2070 ranges from approximately $168 to $212 billion (U.S. Department of Energy, 2000b), comparable to the estimated cost for a 75-year period found in the earlier DOE *Baseline Environmental Management Reports*.

[5] The DOE approach to cost estimation has been repeatedly criticized (U.S. General Accounting Office, 1997a, 1998, 1999). GAO found that cost estimates used in the *Baseline Environmental Management Reports* appear neither to allow for possible future efficiency gains in cleanup technology nor to include costs associated with stewardship for those wastes left in place. Thus, the relative differences between scenario costs may be less than estimated by DOE. These estimates, nevertheless, underscore the sensitivity of total life-cycle costs to the remediation end point selected, as well as changes in planning for types and approaches to remediation.

PURPOSE OF THE STUDY

Official recognition by DOE of the challenges it faces in post-remediation site management is relatively recent. This study by the National Research Council results from one of several recent initiatives directed toward a long-term view of sites in the weapons complex. Some recent studies focused on stewardship are summarized in Appendix D.

The evolution of the DOE remediation programs over time, and the ways that fiscal, technical, and other factors have influenced that evolution, are outlined briefly in the remainder of this chapter. Then a series of questions that were developed by the committee to help frame the committee's conceptual approach is presented. This approach—in essence, the need for reliable, integrated, carefully planned, and iterative long-term institutional management of residually contaminated sites—is laid out briefly here, and in more detail in Chapter 2. Subsequent chapters develop the measures and factors of this conceptual framework as they apply to the management of DOE waste sites. The committee's basic premise is that both capabilities and limitations of DOE to reduce and isolate contaminants as well as to implement and maintain stewardship measures, will need to be taken actively into account in long-term institutional management of waste sites.

The purpose of this study is to identify and examine the long-term challenges that DOE faces in making decisions about waste sites under its control and to suggest an approach to the department's planning and decision making in this arena. The focus of the study was at the individual site level, and did not address such inter-site issues as waste shipments. During this study the committee visited the following sites: Hanford Site, Richland, Washington; Nevada Test Site, Mercury, Nevada; Grand Junction Office Site, Grand Junction, Colorado; several Uranium Mill Tailings Remedial Action (UMTRA) sites in Colorado; Mound Plant, Miamisburg, Ohio; Fernald, Cincinnati, Ohio; Oak Ridge Reservation, Oak Ridge, Tennessee; and Savannah River Site, Aiken, South Carolina (see summary of meetings and visits in Appendix C). During meetings at these sites the committee received and benefited from presentations from, for example, representatives of the DOE Environmental Management Offices of Waste Management, Environmental Remediation, and Long-Term Stewardship, the U.S. Environmental Protection Agency, state regulatory agencies, and from the Environmental Law Institute and Resources for the Future. In addition, during most of its site visits the committee heard comments from local citizens.

Among questions the committee believes that DOE will have to address as site managers begin to develop long-term post-remediation plans are:

- How will the DOE long-term institutional management planning be integrated with current and planned cleanup actions and, where applicable, with ongoing operations?
- To what extent will stewardship measures be relied upon in lieu of site remediation?
- How will the reliability and effectiveness of stewardship, as well as other institutional capabilities, be improved?
- What efforts are to be undertaken to identify and resolve remaining uncertainties regarding the amounts, locations, mobility, and retrievability of contaminants left behind after cleanup is declared to be complete?
- What future findings from long-term monitoring and surveillance programs that are part of stewardship activities will serve to trigger reconsideration of the balance between additional remediation and stewardship?
- What institutions or organizations will carry out stewardship activities, and what incentives will be established to assure that the activities will be carried out?
- What investments should be made in physical and social science and technology research and development so that potentially dangerous contaminants left in place today can be removed or rendered less harmful or more reliably isolated in the future?
- What investments should be made in research and development to improve the likelihood that long-term institutional management measures will remain efficacious over the long term?
- How can adequate and reliable funding be assured for all of the activities that are necessary parts of an effective long-term institutional management program?

At a more general level, the questions to be faced also include:

- What approaches will be taken by the long-term management organizations that emerge—what goals will they have and what goal-setting approaches will they use?
- What ability will these organizations have to operate in a systems-oriented, comprehensive, and integrated way?
- What will be the ultimate character of the institutions and organizations charged with the care of contaminated sites (e.g., funding mechanisms, accountability, ability to detect and correct errors)?

These and related questions that emerge from this study are developed and explored in subsequent chapters. Though none is easily answered, many are now beginning to receive attention within DOE. Nevertheless, confronting the "Faustian bargain" referred to by Alvin Weinberg will require both a high level of commitment today and a level of vigilance into the distant future that is unusual, if not unprecedented. This vigilance should not be expected to materialize spontaneously. Thus far, policy debates have not adequately considered the magnitude of the challenges that need to be faced.

LONG-TERM INSTITUTIONAL MANAGEMENT

In this report the committee will describe and discuss the term *long-term institutional management*, referring to the measures over long periods of time used to ensure that public and worker health and safety and the environment are protected when potentially hazardous contaminants are left on sites. The degree and scope of protection needed will depend on the nature of the residual contaminants and the possibilities for exposure to them, both now and in the future. Thus, the measures will also depend on how the sites are used, now and in the future. The time frames over which institutional management measures must be effective are set by the lengths of time over which residual contaminants can be expected to pose unacceptable risks. This last aspect of long-term institutional management is notable because some radionuclides and other contaminants that will be left at sites can be expected to remain as risks to the public and the environment for thousands of years (U.S. Department of Energy, 1999). However, among the current most hazardous radioactive materials with respect to risk that are found in the DOE defense complex are the fission products found in high-level waste (such as cesium-137 and strontium-90). These radionuclides, currently stored in tanks, drums, capsules, trapped within processing facilities, or contaminating the soil and groundwater where they have been released or leaked, must be very carefully managed for at least the next 300 years, a period extending beyond the projected closure date for most sites by DOE. In addition to these relatively short-lived isotopes, there are quantities of very long-lived isotopes such as the transuranic elements that may require vigilant care much farther into the future.[6]

The committee's conceptual framework for examining the requirements for successful long-term institutional management, first discussed in its interim report (National Research Council, 1998d), is presented in Chapter 2, followed by discussion of the basic measures of such a form of management:

- contaminant reduction (Chapter 3),
- contaminant isolation (Chapter 4), and
- stewardship activities (Chapter 5).

---

[6] Fission product radionuclides now consist mainly of cesium-137 and strontium-90 with half-lives of about 30 years. Thus, natural attenuation (decay) will reduce contamination by a factor of about 1,000 in 300 years. Transuranic and other long-lived radionuclides and many other hazardous chemicals, on the other hand, can be expected to remain for thousands of years—essentially forever in the case of most stable heavy-metal contaminants.

There are, in addition, numerous contextual factors or elements that affect long-term site disposition, as is discussed in Chapter 6. In Chapter 7 we discuss the fact that technical and institutional systems have limited capabilities, and these capabilities and limitations will directly affect management decisions and activities. These considerations taken together frame the problem of how best to approach the design and implementation of durable and effective institutional management systems. Chapter 8 provides some design principles for institutional management and the committee's findings and recommendations.

Throughout this report the committee has presented long-term institutional management characteristics and principles. These are introduced in Table 1 to assist the reader, with indication of where the terms are defined and discussed.

TABLE 1   Institutional Management Characteristics, Criteria, and Principles Found in This Report

A. *Characteristics of an Effective Stewardship Program (from Chapter 5)*
- Layering and redundancy.
- Ease of implementation.
- Monitoring commensurate with risks.
- Oversight and enforcement commensurate with risks.
- Appropriate incentive structures.
- Adequate funding.
- Durability or replaceability.

B. *Characteristics of Institutional Design (Design Criteria) (from Chapter 8)*
- Defense in depth.
- Complementarity and consistency.
- Foresight.
- Feasibility.
- Accountability.
- Transparency.
- Stability through time.
- Iteration.
- Follow-through and flexibility.

C. *Five Key Principles for Developing an Institutional Management System (from Chapter 8)*
- Plan for uncertainty.
- Plan for fallibility.
- Develop substantive incentive structures.
- Undertake scientific, technical, and social research and development.
- Seek to maximize follow-through on phased, iterative, and adaptive long-term approaches.

D. *Characteristics of Implementation Criteria (from Chapters 5 through 7)*
- Clear objectives and a desire on the part of those responsible for institutional management to carry out those objectives with diligence over time.
- A clear system of governance that specifies what is to be done and by whom and is founded on precepts that are enduring on the one hand and flexible on the other.
- An integrated overall approach that coordinates activities across the responsible entities and assures that site management measures are complementary rather than conflicting.
- Incentives both within and outside the institutional management organization to encourage diligence in carrying out mission objectives.

2

Conceptual Framework

Any sites retaining hazardous contaminants over a long time period will require specific forms of dedicated and ongoing vigilance. To focus and systematize its review of the major challenges that the U.S. Department of Energy (DOE) will face as sites undergo the transition from mission-oriented operations to remediation and closure as described in Chapter 1, the committee devoted the initial period of its study to developing a conceptual framework for long-term site disposition decisions. This chapter presents an overview of the committee's conceptual framework—*long-term institutional management*. The emphasis is on sites that face the prospect of continued management over very long periods of time.

GENERAL REQUIREMENTS

The committee's conceptual framework embodies the considerations that must be taken into account for planning remediation and stewardship activities at individual sites. *In all cases reviewed by the committee, current DOE remediation planning and planning for post-remediation site stewardship can fit within the conceptual framework. In no case reviewed, however, was planning and management developed to a degree that the committee's framework suggests it should be.*

Long-term institutional management of contaminated sites should be:

- *realistic* in being based on recognition of practical constraints as well as capabilities;
- *systematic* in its overall approach; and
- *integrative* and *comprehensive* in its consideration of three measures:
    1) the types of *contaminant reduction* measures employed;
    2) the types of *contaminant isolation* measures employed; and
    3) the reliance placed on *stewardship* measures,

so that the balance achieved among reliance on each of these three types of measures is appropriate given the following contextual factors:

- risks to members of the public, workers and the environment;
- technical and institutional capabilities and limitations and the current state of scientific knowledge;

- costs and related budgetary considerations;
- legal and regulatory requirements;
- values and preferences of interested and affected parties; and
- impacts on other sites.

Each of these contextual factors will receive different emphasis at different sites, depending on the site characteristics and the surrounding areas and populations, as well as social, economic, legal, and political considerations at both the local and national levels. The application of long-term institutional management should also be:

- *phased* and *iterative* in its execution over time, with goals that are themselves adjusted over time in response to changing knowledge and public opinion, and the establishment of responsible and authoritative organizations to ensure that the goals are met;
- *adaptive* in the face of future opportunities or challenges to improve upon imperfect solutions imposed by technological and other constraints; and
- *active* in its search for knowledge that reduces uncertainties and in seeking the technical, institutional, and financial means to improve upon past decisions.

The committee's conceptual framework for long-term institutional management is the metaphor of a ***three-legged stool***, with its legs corresponding to the three measures mentioned above—contaminant reduction, contaminant isolation, and stewardship. This metaphor is useful for two principal reasons. First, it highlights the measures that must be present for the metaphorical stool, viewed as a "system," to be complete and stable. Second, it emphasizes the interrelationships among these measures that are necessary to maintain that integrity over time and to give the stool the overall character that it needs, given the environment in which it will be used.

The three-legged stool that symbolizes long-term institutional management of contaminated sites is illustrated in Figure 2. The stool's *legs* symbolize the principal measures available to managers making disposition decisions aimed at site completion in the sense defined in Chapter 1. ***Contamination reduction measures*** (Chapter 3) are actions taken to reduce the amount of contamination by removal or in situ destruction (e.g., bioremediation). ***Contamination isolation measures*** (Chapter 4) are engineered measures implemented to stabilize, fix, or impede release of or access to contamination at a site. Physical barriers, and chemical or thermal fixation of wastes, are included in this category, as are "pump-and-treat" (or "pump-and-reinject") actions aimed at retarding migration of subsurface plumes. Natural attenuation (including radioactive decay) is also included. Remedial action measures are, then, any combination of contaminant reduction and contaminant isolation measures. ***Stewardship measures*** (Chapter 5) include measures to maintain contaminant isolation and reduction technologies and to monitor the migration and attenuation of residual contaminants, as well as such measures as land use and access restrictions (institutional controls), oversight and enforcement, information management, and periodic reevaluation of protective systems. The latter include consideration and use of new technological options to reduce, eliminate, or contain residual contaminants.

Like any metaphor, the three-legged stool as a metaphor for long-term site management is not perfect. It will not apply, for example, in situations where contaminant reduction is sufficient to allow unrestricted use or if engineered contaminant isolation measures are not required. At complicated sites, however, some reliance on all three of the functions represented by the stool's legs—contaminant reduction, contaminant isolation, and stewardship measures—can usually be expected. Also, the three measures are not as independent of one another as the figure suggests, as stewardship, properly construed, applies to each of the other two legs.

In many cases, the three sets of measures will bear a "funnel" relationship to one another. Remediation occurs first, then containment barriers are applied to the residual that remains, and then institutional controls are developed to protect humans and the environment from harm. The three-legged stool metaphor emphasizes the dependency on all three measures at the expense of indicating the actual order of application in some instances. In doing so, however, it serves to make a key point that stewardship is a pervasive concept and not simply a set of measures to be implemented once remediation is complete.

The emphasis placed on each of these three basic measures in site disposition decisions will depend on the

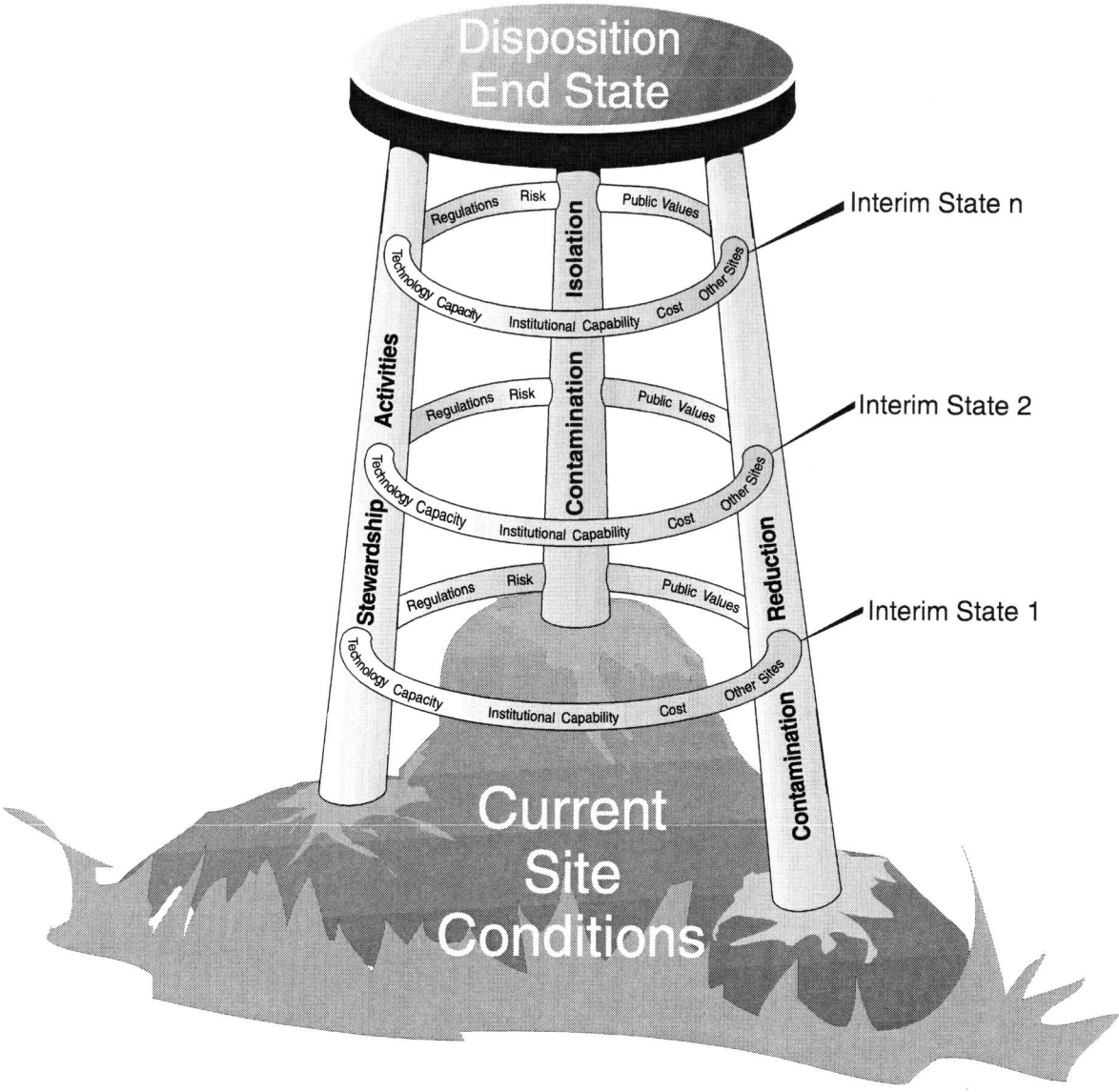

FIGURE 2 Long-term institutional management conceptual framework.

capabilities and limitations of each, as well as on the nature of the site problem that is being addressed (i.e., metaphorically, the *terrain* upon which the stool is to rest), and in part on broader contextual factors (such as risks, costs, public values, legal and regulatory requirements, technical and institutional capabilities and scientific knowledge, and impacts on other sites). The latter are symbolized by the *rungs* of the stool, which, metaphorically speaking, connect and fix the legs. The role that such factors play in long-term disposition decisions is elaborated upon in Chapter 6. The rugged terrain on which the stool rests (illustrated in Figure 2) is intended to represent the wide range of contamination characteristics present between and within sites that will drive decisions toward determining the appropriate balance of reliance to be placed on contaminant reduction, contaminant isolation, and stewardship measures, given site characteristics.

The legs of the stool in Figure 2 support a ***seat*** that symbolizes a planned end state, which may or may not be the final goal envisioned for the site. Here again, the limitations of the metaphor must be recognized. In some ways the seat is the ultimate purpose of the stool—it is both an end in itself and the means by which desired future site uses can be achieved—but, at the same time, the distinctly limited capabilities to anticipate changes at a site, or in society and its technologies more broadly, must be acknowledged. For example, the range of possible future land uses may broaden as remediation technologies improve. On the other hand, the range of potential land uses may narrow, or unsuspecting future citizens could be exposed to unacceptable risks, if contaminant isolation or stewardship measures begin to fail.

In the conceptual model, progress toward planned goals occurs in stages. This iterative, phased feature of long-term disposition decision making is symbolized by the successive layers of rungs (or *interim states*) in the stool pictured in Figure 2. Interim cleanup goals are currently in wide use throughout the DOE complex, and even where end state goals have been selected, they may have been set provisionally. Or, the remedial actions necessary to achieve them may need to unfold in successive stages over fairly long periods of time. As will be discussed in Chapter 7, there is a clear need to recognize that, at present, we do not have reasonable assurance that such follow-through can reliably be counted on to occur. Successive stages of evaluation could involve reconsideration of the goals previously selected, or adjustment of how the three sets of measures symbolized by the stool's legs are to be applied to attain the selected goals. Finally, the nature and relative importance of the individual contextual factors that make up the rungs, and the interrelationships among these factors, can also change through time.

SITE DISPOSITION DECISIONS FROM A LONG-TERM INSTITUTIONAL MANAGEMENT PERSPECTIVE

The sites and situations that the committee considered are extremely varied in character. Neither within the DOE complex nor with regard to contaminated industrial sites in the private sector do many generalizations apply to all waste site disposition decisions. This section elaborates on selected aspects of the committee's long-term institutional management framework as applied to current practice or planning at individual sites.

"End States" as Guides to Site Disposition Decisions

The iterative or phased character of remediation efforts, with goals successively defined and redefined, was apparent at many of the DOE sites visited by the committee. Although individual remediation actions are usually directed at relatively well-defined end states (typically, cleanup goals set by the U.S. Environmental Protection Agency [EPA] or state regulators for groundwater or soils), the ultimate end state for the site as a whole may for all intents and purposes be unknown, and may remain so for a considerable time as site remediation proceeds. Especially for the larger sites, *end states appear at present to be emerging as the de facto result of multiple interim actions*. These interim actions are aimed at achieving interim states and are being applied in serial fashion via regulatory definition to often relatively small and relatively dispersed, former operational units within the larger site, or facilities or disposal areas within operational units.

The effect of cleanup proceeding this way is to produce a relatively clean site with pockets that may remain contaminated and therefore in need of institutional management into the indefinite future. The larger sites within the DOE defense complex appear to be evolving toward a "Swiss cheese" configuration, which while potentially able to support multiple uses in land areas where successful contaminant reduction or the lack of contamination in the first place enables unrestricted use, may also present challenges for ongoing management efforts in other areas where stewardship measures are required because residual contamination persists and represents a hazard.

One example of site cleanup proceeding this way is the program of soils cleanup at the Hanford Site 100 Area, adjacent to the Columbia River. Cleanup is guided by highly specific target decontamination levels tailored to the risk to a hypothetical future resident atop the filled pits and trenches that are small parts of defined operational units. Disposition of reactors, spent fuel sites, buried wastes, and other nearby contaminated sites is proceeding on separate tracks and time scales, with relatively few final decisions yet made about the specifics and ultimate goals of remediation efforts (particularly with regard to the ultimate disposition of reactors and contaminated groundwater plumes).

This fragmented decision-making approach appears to have developed for a variety of reasons. Under the Comprehensive Environmental Response, Compensation, and Liability Act of 1980, as amended (CERCLA) and the Resource Conservation and Recovery Act of 1976 (RCRA), the dominant federal laws governing cleanup at most DOE sites, cleanup goals are likely to be negotiated over time on an operational unit or solid waste management unit basis. (Appendix E contains a summary of the legal structure for closure of DOE sites.) When wastes are left on site, as is likely at most large sites, all operable units may not be able to achieve the same cleanup goals. For example, the presence of DNAPLs (dense, non-aqueous phase liquids), metals, or other difficult-to-address contamination problems may mean that cleanup levels (i.e., ARARs—applicable or relevant and appropriate requirements—under CERCLA) may not be achievable and a finding of technical impracticability may be made under CERCLA (National Research Council, 1999e). As demonstrated by the Hanford Site 100 Area, cleanup will tend to move forward on problems where agreement exists, while decisions in other areas of the larger cleanup problem are deferred.[1]

The ultimate disposition of contaminated facilities within larger areas undergoing remediation can also strongly influence the end state that is achievable for the larger area. An example includes the plutonium production reactors in the 100 Area at the Hanford Site where the C Reactor has recently been put into "interim safe storage" by placing it in a "cocoon" that has a 75-year design life (see Sidebar 2-1). But major decisions remain to be made on the disposition of the reactors in the 100 Area as a group. Current options range from permanent entombment in place (following removal of the reactor cores to permanent disposal in the 200 Area, the Hanford Site's central waste management area) to physical removal of entire reactor buildings to this same area. One variant has the historic B Reactor remaining on site to become a museum, open for public visits.

Even where future land use preferences guide the choice of remediation end points, the experience of the EPA Superfund program suggests that the correlation between land use preferences and end point selection is poor (Hersh et al., 1997). Thus, *exit point* from a particular phase in site management is perhaps a more accurate term than *end state* to define site condition at the point where remediation ends. An absence of the kind of systems-oriented thinking that is espoused in the committee's long-term institutional management framework is evident in these examples drawn from the Hanford Site. A systems engineering approach for analyzing various pathways with related uncertainties toward an end point has been discussed in several recent reports about cleanup of Hanford from the National Research Council (1998a, 1999d).

Tradeoffs Between Present-Day Remediation and the Need for Long-Term Stewardship

For numerous contamination problems within the DOE complex, site disposition decisions rely heavily on engineered containment and subsequent stewardship activities. The situations are quite varied and include cases where fairly extensive waste removal is being undertaken, as well as those in which relatively little waste is being removed. Examples in the latter category include natural attenuation sites and sites receiving "technical impracticability" waivers under EPA guidelines (National Research Council, 1999e).[2] While technical impracticability waivers have to date been applied on a very contaminant- and situation-specific basis, these cases will nevertheless

---

[1] Although there is strong local support for the approach being taken to the Hanford Site 100 Area soils cleanup, the use to which these lands along the Columbia River are ultimately to be put remains undecided. The DOE Inspector General (IG) has argued for less stringent cleanup standards than those in use in the soils cleanup, stating that residential standards are inappropriate to the future land uses that are most likely to be adopted. The IG estimates that a cost savings of $12 million would result from the remediation of just the first few of the 70 or so soil sites if they are cleaned up to a "rural-residential" land use scenario (*Seattle Post-Intelligencer*, July 8, 1999). Under current plans a total of some 3 million cubic yards of soil and solid waste will be excavated (U.S. Department of Energy, 1999, Appendix E).

[2] EPA's "technical impracticability" waiver guidance is aimed at DNAPLS and similar difficult-to-remedy groundwater cleanup problems found at Superfund sites. DOE's contention that some cleanup problems around the complex (e.g., the underground test cavities at the Nevada Test Site [see Appendix F], the hydrofracture zones at the Oak Ridge Site) are technically impractical to clean up does not constitute a regulatory determination to that effect. There is no formal process for making such declarations with respect to DOE sites, other than where the contaminants are similar to those for which such declarations have been made at privately owned sites.

SIDEBAR 2-1

HANFORD SITE REACTOR 'INTERIM SAFE STORAGE'

In the Hanford 100 Areas, one reactor, the C Reactor, has been put into an "interim safe storage" condition (Richland Environmental Restoration Project and Bechtel Hanford, Inc., 1999). Construction on this water-cooled, graphite-moderated production reactor was begun in June 1951 and it was started up in November 1952, just 17 months after groundbreaking. It was one of nine constructed between 1942 and 1955 at the Hanford Site along the Columbia River to produce weapons-grade plutonium. The reactor was shut down in April 1969, and deactivation was completed in early 1971.

The C Reactor building was 106 m by 93 m (346 feet by 305 feet) in size, with a height of 30 m (98 feet) and constructed of reinforced concrete in its lower levels and the central portions surrounding the reactor. The design objectives for the interim safe storage included:

- safe storage for up to 75 years;
- no releases of radionuclides to the environment under normal design conditions;
- required interim inspections on a 5-year frequency basis; and
- completion of a safe storage enclosure configuration that would not preclude or significantly increase the cost of any final decommissioning alternative.

The safe storage condition for the reactor building included several significant steps. A significant portion of the structure outside of the reactor was removed (reducing its area by about 80 percent of its original size). Before this occurred, the highly contaminated sediments from the irradiated fuel element discharge area that were stored in the fuel storage basin transfer pits were encapsulated in grout to form monoliths. Finally, the remaining reactor core was encased in 3- to 5-foot thick concrete shielding walls and a corrosion resistant galvanized steel roof. Although initial planning for the C Reactor decontamination and decommissioning project included filling the main reactor building with grout, this element of safe storage enclosure construction was abandoned out of concern that grouting might preclude later dismantling of the entire reactor structure and moving it to the waste management area on the site's central plateau (the 200 Areas). The safe storage enclosure was completed in September 1998.

Major decisions remain to be made on the disposition of the production reactors as a group, with current options ranging from permanent entombment in place to physical removal of the main reactor buildings in whole or in pieces to the Hanford Site's central waste management area (the 200 Areas). It has been suggested that reactor building removal could be accomplished via tracked vehicles similar to those used to move the Space Shuttle to its launch pad. An interesting variant on all these options has the B Reactor remaining on site to become a museum to the Atomic Age, open for public visitation. The consequences of this latter possibility for current cleanup planning for nearby lands within the 100 Areas have as yet received scant attention.

REFERENCE

Richland Environmental Restoration Project and Bechtel Hanford, Inc. 1999 (February). Submittal for 1998 Project of the Year—C Reactor Interim Safe Storage. Richland, Wash.

---

often necessitate that humans and the environment be protected from contact with contaminants for very long periods of time.

Where physical systems like pump-and-treat are employed, they may need to be maintained in good working order for very long time periods, placing additional burdens on site stewardship. At Hanford, some 85 square miles of the site are underlain by contaminated groundwater that currently does not meet drinking water standards (U.S.

Department of Energy, 1999, Appendix E). Groundwater plumes are contaminated with radionuclides and other hazardous chemicals and are now impinging on the Columbia River. A pump-and-reinject system of wells is in use in the 100 Area in an attempt to retard migration of strontium into the river. The intention is that natural decay will reduce radiation levels to drinking water standards (8 pCi/l for strontium) before significant release to the river occurs. With strontium-90 having a half-life of 29 years, such an operation may have to be run for about 300 years to reduce the radioactivity by a factor of about 1,000.

Choices between reduction of contaminants now and continued reliance on contaminant isolation and stewardship measures far into the future exist throughout the complex. At the Nevada Test Site (NTS), for example, fairly extensive surface soils cleanup has been directed at plutonium dust and fragments (the result of subcritical nuclear testing), while no remediation is contemplated for the underground test cavities (Appendix F). DOE has attempted to set goals for cleanup at NTS that it believes are consistent with the site's anticipated future use as a high-security standby site for the possible resumption of underground nuclear testing and its relative remoteness from human populations. Some commercial use of the NTS is also contemplated, possibly including a satellite launching center.

Adaptability and Flexibility in Remediation Approaches

At DOE sites for which no currently known waste removal option exists, the long-term nature of the problem poses a dilemma. The inability to foresee future land use, possible failure of containment barriers or other remediation technologies or development of better ones, or the character of future society, are all factors that point to the need for building adaptability and flexibility into current site remediation planning. Adaptive and flexible approaches can take a wide variety of forms (for example, the Hanford Site reactor "interim safe storage", see Sidebar 2-1). The Hanford decision to abandon the use of grout vaults for on-site disposal of the low-activity fraction of the wastes separated and removed from the high-level waste stored in underground tanks in the 200 Area was based in part on similar considerations. This decision shifts from disposal of these wastes in the 200 Areas in the form of grout vaults to the form of containers of vitrified waste that can be stored in a variety of locations, albeit with their own inherent problems.

In summary, long-term institutional management is a concept that represents a systematic approach to protect the public and the environment from contaminants that remain at sites upon cessation of remediation activities. It includes three sets of measures that are supported by applying the results of the new scientific understanding and technical development:

1. contaminant reduction—actions that may be applied to reduce the level of risk presented by the residual contaminants;

2. contaminant isolation—actions taken to monitor existing barriers to residual contaminant migration and to reduce the chance of migration in the future; and

3. stewardship—actions taken by responsible authorities to protect the public and the environment from risks present at residually contaminated sites.

Although these three sets of measures may be implemented sequentially, planning and decision making for them must be conducted simultaneously, based on the existing conditions and the desired end point. Affecting these measures are a number of contextual factors, many of which address the uncertainty of present and future capabilities and limitations. These three sets of measures and the contextual factors will be discussed in greater detail throughout this report. The committee uses terms relevant to institutional management, as described in this chapter, throughout the report: definitions of the terms are listed in Appendix I.

3

Contaminant Reduction

As mentioned in Chapter 2, contaminant reduction is one of the three sets of waste-site measures embedded in the long-term institutional management approach. The focus of this chapter is the existing contamination at U.S. Department of Energy (DOE) sites and the goals, constraints, and limitations for its remediation via contaminant reduction. The role of scientific and technology research in improving contaminant reduction is also discussed.

Contamination at a site may be reduced in volume and toxicity by processes such as destruction, decontamination, processing to form a more concentrated waste stream and a less hazardous secondary waste[1] product, transmutation to a less hazardous form, decay of radionuclides or of certain other hazardous substances, and removal from the site. Examples of destruction might be the incineration of certain substances or contaminated materials, biodegradation treatment, or in situ vitrification of contaminated materials, such as soil, that may destroy certain organic materials and immobilize others. These techniques require collection and handling of any hazardous residues such as gases and ashes from the destructive process. Decontamination of buildings and other structures is being conducted at many DOE sites, often resulting in a reduced amount of contaminated material, albeit more concentrated, for further management as well as some potentially useful materials and structures.

Another form of contaminant reduction includes removal of the mobile species from soil by pumping or vapor vacuum extraction and the collection of such materials for disposal or recycling for future use. Processing of high-level radioactive waste from tanks may be used to separate and concentrate the more radioactive materials for long-term management, leaving behind a lower activity waste that may have a less stringent requirement for future management. Some proposed solutions for the safe management and disposal of highly radioactive wastes containing long-lived radionuclides have focused on separating these radionuclide components of the wastes and transmuting them by neutron bombardment to form nuclides that would be either stable or radioactive with much shorter half-lives (National Research Council, 1996b). (However, development of transmutation is at such an early stage that it holds little hope for treating contamination at the sites.) Finally, the contaminants may be recovered from a site, placed in some type of acceptable waste form for transport and internment, and moved to another site, resulting in a transfer from one location to another with the expectation that the wastes will be confined in a manner that presents less risk to the public and the environment.

---

[1] Secondary waste is new waste produced in the course of carrying out the processing, concentration, and removal of contaminants.

At the present time DOE legacy waste site managers are placing most attention and funding on the reduction of contamination rather than isolation or stewardship. The radioisotopes are in wastes classified as high-level, low-level, spent fuel, transuranic (TRU), or mixed (hazardous and radioactive materials together). The waste is present in tanks of varying sizes from several hundred gallons up to over 1 million gallons (3,800 m^3), in burial pits from small to very large in volume, in wet and dry storage canisters, in drums and other packages, and stabilized in some forms such as glass or grout. It is also found where it has been purposefully or accidentally disposed of in the soil and the groundwater, and adhering to or contained within buildings, machines, scrap metal, concrete, protective clothing, cleanup substances, and other materials that were involved in the generation, processing, and storage of nuclear materials and the production of nuclear weapons.

There are a number of non-radioactive, hazardous substances of concern at many, if not all, facilities within the DOE complex. These include both elements (especially metals) and compounds that never degrade and organic compounds that can degrade. They range from hazardous substances remaining in cleaned-up tanks to contaminants in soils and groundwater to contaminated surfaces. Hazardous substances found in remediated tanks include residual waste, lead used as shielding, and chemicals used in cleaning, plating, reprocessing, and separations operations as well as in machining and fabrication operations. The decontamination approaches for these materials are often the same as those used for radioactive materials, although the regulations governing them and the permissible ultimate disposal methods may be very different.

An extensive program of contamination reduction (decontamination or destruction) of radioactively contaminated equipment, facilities, buildings, groundwater, and soil at the DOE sites is planned in connection with the long-term disposition of DOE legacy wastes (U.S. Department of Energy, 1997b, 1998a). Decontamination of a site is usually the first step in site remediation. Typically the goal of decontamination is to produce two streams: 1) a product stream—a site, facility, or piece of equipment suitable for some sort of beneficial use, or at least posing a reduced risk, and 2) a waste stream that contains the contamination. The contaminated site, facility, or equipment is treated by physical, chemical, or biological means to achieve this end. If the degree of decontamination reached is insufficient to permit the required separation and removal of the contamination, the contaminated material must be isolated from the biosphere (see Chapter 4).

Examination of the table in Appendix B showing closure plans for major DOE sites reveals that almost none of the significant DOE sites will be cleaned to residential/agricultural standards in all their parts. Rather, most will be cleaned to a mixture of cleanup levels ranging from residential/agricultural to controlled access. Also, many sites will have continuing missions, with only parts of them to be made available for other than DOE uses. Uranium Mill Tailings Radiation Control Act of 1978 (UMTRCA) disposal sites form a class by themselves. A negligible number of sites revert to the original owner after cleanup; some will be turned over to state authorities. Practically all leave management responsibility of residual contamination, if any, to DOE. Approximately 19 sites from Appendix B have some degree of nongovernmental ownership. Some sites are designated to be monitored or have open-ended pump-and-treat requirements for unspecified periods of time. Management of some sites dictates both indefinite monitoring and pump-and-treat.

Some observations and conclusions that may be derived from the table in Appendix B are:

- Many of the sites are to be released all or in part for restricted use.
- Some of the major sites are to be released in part for unrestricted use.
- Many of the major sites have ongoing DOE missions into the unspecified future.
- Many of the sites, major and other, are subject to open-ended pump-and-treat remediation (i.e., pump-and-treat is the method of choice for many long-term groundwater problems).
- Robust institutional management will probably be needed for a majority of the sites.

In fact, one might reasonably argue from the information in the table that there is a need for new, imaginative, and practical follow-on or alternative groundwater cleanup or isolation methods (a science and technology focus), that there is a need for effective, long-term on-site and off-site monitoring methods, and that responsibility for those using sites after closure (e.g., federal, state, or local government agencies or the private sector) must be clearly identified and formalized.

Large amounts of radioactive metal from equipment, utilities, and structures need to be decontaminated, along with large amounts of radionuclide and chemically contaminated soil and groundwater that need to be decontaminated or dealt with in some other manner. Ideally the result of decontamination should be a site, facility, material, or equipment that is free from any restrictions on its use, and the waste stream should be small, well characterized, and economical to produce and dispose of. This ideal is difficult to achieve, however, and in general will only be reached in favorable cases. In view of the range in the levels of contamination and the sizes and number of the contaminated sites and facilities, it is expected that the levels of decontamination likely to be achieved will range from essentially complete decontamination, allowing the site to be released for unrestricted use, to levels of decontamination that require long-term institutional oversight, control, and monitoring. The possibility of additional remediation to further reduce the risk from a contaminated site will remain until contamination no longer poses a hazard.

A necessary adjunct to promote success in contamination reduction is planning for contingencies. It is quite possible that, for reasons such as inadequate knowledge of the contaminated facilities or environment, or lack of appropriate and tested technology, a planned decontamination operation will fail to achieve the desired or agreed upon future state, or that future state may change. To allow for such a contingency, one or more promising alternative approaches should be identified, developed, tested, and available to ensure that the risk posed by the contamination is managed acceptably. One such alternative might be to go to isolation of the contamination, as discussed in Chapter 4.

FUTURE STATES

Decontamination is one of the essential aspects of site remediation that can lead to an agreed-upon future state. A future state need not necessarily be the final or end state (or condition) of the object or site being decontaminated. In essence, there may be interim states such that waste management may be phased, and that additional contamination reduction may be carried out to reduce risk in the future on sites or objects remediated to interim states. Such possibilities will continue to exist if a dynamic program of scientific and technical development is pursued toward improved scientific understanding and new methods for contamination reduction, even after a site or facility has been deactivated.

Ideally, for all the necessary operations to be successfully carried out to attain the agreed-upon future state, such a state should be defined in advance of the contamination reduction operations. The future state is usually defined in consultation with and by agreements with regulatory bodies and other parties having a legitimate interest in site disposition. For interim states there should be future reevaluations, presumably at agreed-upon intervals, to determine if technological, regulatory, or institutional changes make further reductions in contamination desirable and practicable, and if so, to see that they are carried out. It is also possible that decontamination and cleanup standards or goals may have changed. This concept is elaborated in Chapter 6 of this report.

The objective of the contaminant reduction operations, both in the current and interim state, is removal or destruction of the source of contamination to the extent possible, reducing reliance on containment and stewardship activities while achieving better future conditions. However, as a general rule, the greater the degree of decontamination, the greater the cost, and in some cases the greater the worker risk, the contaminant by-products, and the environmental disturbance. In practice, a balance should be sought between the degree of decontamination and the fiscal and health risks and the environmental insults associated with cleanup and the waste streams it will create. Standards for achieving sufficient decontamination are very important. There is a significant lack of clear standards for unrestricted release of decontaminated sites.

The goal of decontamination may also be to move a contaminant to a location where it poses less threat to the public and the environment than it did in its pre-decontamination site. Thus, decontamination can also result in a wider range of possible future and end states for some sites. It should also be recognized that there will often be a trade-off between decontamination and containment (discussed in Chapter 4). In some cases there may be a cost-benefit advantage to containing part or all of the contamination rather than removing or destroying it. Similarly, until improved decontamination technologies are developed, containment may entail less risk to workers and the environment than contamination reduction. For example, dredging of contaminated sediments from streambeds may cause an unacceptable increase in exposure to workers as well as to the public and the environment.

Different interim and end states may be possible in the future due to (a) development of new technologies and more economical cleanup to lower levels of residual contamination; (b) availability of additional resources; (c) changes in the values of the interested and affected public and regulators; or (d) failure of the remediation approach used or of stewardship measures. A phased approach (one that proceeds toward a goal in stages while important information and technology gaps are filled) to contaminant reduction and final disposition of a still-contaminated facility or site will allow for future resolution of current unknowns and uncertainties and for new technologies and methodologies (National Research Council, 1996d).

A large number of decontamination technologies is available today, as discussed in recent reports of the National Research Council (1994c, 1996e, 1997b, 1998b, 1999c,e). The preferred ones are likely to be those that produce the least amount of secondary waste, are the most economical to use, and provide the lowest risks to the workers, the surrounding community, and the environment during the decontamination operations. The many regulations governing acceptable levels of decontamination for various purposes are discussed in Appendix E. A significant problem in these regulations is the absence of volumetric standards.

Waste Storage Tanks

Underground waste tanks at the Hanford Site in Washington, the Savannah River Site in South Carolina, the Oak Ridge Reservation in Tennessee, and the Idaho National Engineering and Environmental Laboratory were used for storage of liquid waste from the processing of irradiated fuel elements. In general, sites with underground tanks formerly used for high-level radioactive waste will not be released for public use in any foreseeable time frame. The degree of decontamination achievable is not known, and will doubtless differ from one tank to another. It is not clear what the trade-offs will be between contaminant reduction and containment. Also, from what the committee has learned, it is not clear what the final criteria will be for tank cleanup and closure. Closure measures required for waste tanks at the major DOE weapons sites are typically viewed as a matter that should involve the interested and affected public, DOE, the states, the U.S. Environmental Protection Agency (EPA), and Native American tribes. The public is broadly viewed as including local residents, health organizations and environmental activist groups, and others not directly associated with the site.

However, the type and degree of decontamination required for tanks are not entirely matters of agreements, cost, and risk. Very definite limitations are imposed by the physical nature and condition of the tanks and by the state of the art of tank decontamination technology. For example, most tanks were built without consideration of their final disposition, and many have been in use beyond their planned lifetimes. They often have dozens of internal structures for purposes such as transfer of contents, monitoring systems, structural reinforcement, venting, cooling, and sampling. These features are significant impediments to removal of sludges and solids that lie in many, if not most, of the large tanks (especially those at the Hanford Site). They not only present a significant problem in carbon steel tanks containing neutralized waste, but they also can inhibit extraction of acid wastes from stainless steel tanks.

The appropriate degree of waste extraction from tanks has been the subject of extensive discussion between DOE and the U.S. Nuclear Regulatory Commission (USNRC) for the past 10 years. Early discussions concentrated on identifying whether wastes could be sufficiently identified as high-level by identifying their source. This led to efforts to change the definition of high-level waste to cover a greater amount of wastes. In denying a petition to change the definition of high-level waste, the USNRC gave three criteria to be used to determine whether high-level waste has been extracted and waste incidental to reprocessing remains (Bernero, 1993). Those criteria are: "[the waste] (1) has been processed (or will be further processed) to remove key radionuclides to the maximum extent that is technically and economically practical; (2) will be incorporated in a solid physical form at a concentration that does not exceed the applicable concentration limits for Class C low-level waste as set out in 10 CFR Part 61; and (3) will be managed, pursuant to the Atomic Energy Act, so that safety requirements comparable to the performance objectives set out in 10 CFR Part 61 are satisfied." These criteria offered DOE a reasonable basis to judge when the bulk of the high-level waste has been extracted and the residues, still in the tanks or from the high-level waste concentration process, can fairly be classified as waste incidental to reprocessing. The DOE has promulgated its overall criteria for the management of radioactive wastes in DOE Order 435.1 and its supporting documents.

Recent experience with waste retrieval from tanks is revealing problems that have implications for the long-term institutional management of the tanks once wastes are removed. Historical records are used to estimate the initial load of waste in the tank because an assay of the residue can be extremely difficult. Some of the residue is sludge or solids trapped in difficult-to-reach locations of the structures; some is caught on or in the corroded carbon steel surface of the tank interior. Thorough, direct sampling and characterization are not really possible in a practical sense. Current tank cleaning and extraction are being done with water-based hydraulic methods. There are questions about the in-tank residue and whether it can meet Class C limits through a concentration-averaging approach. More aggressive techniques, such as using acid flushes, may not turn out to be technically or economically practical and may even cause leaks in the tanks in the attempts to remove the wastes. It is apparent that many of the tanks' physical structures will be left in place following removal of most of the waste.

The current state of the art leaves two evident options for the tank closure process. If the USNRC criteria for high-level waste removal are used to judge what is acceptable, residual wastes in the tanks would be acceptable if they do not exceed Class C waste performance standards. If this type of in situ disposal is accepted, this option leads to a final stabilization solution (end state) that involves filling and surrounding the tanks with a grout or with concrete engineered barriers (see Chapter 4). Site monitoring would still be required. In contrast, an interim tank closure might be selected in anticipation of development of more effective treatment technologies for the tank and its residues at a future time. In this approach, the tanks and their contents would be stabilized in a reversible way (e.g., by filling the tanks with gravel or "poor grout" [grout that is removable] and establishing a pump-and-treat system to recover any wastes that might escape). Most of the tank wastes contain currently hazardous fission product radionuclides having half-lives of about 30 years or less (e.g., strontium, cesium, and tritium); these radionuclides will decay by a factor of at least 1,000 over about 300 years. An incentive for adopting the interim closure option with its greater burden of institutional management would be, where warranted, the flexibility it provides for more complete tank decontamination at some future date if improved or new decontamination processes are developed.

Buried Waste

A certain amount of long-lived radioactive waste currently buried will probably be removed for further processing and transfer to another storage site or to a repository. Such waste includes TRU wastes buried in trenches and pits in the Radioactive Waste Management Complex at the Idaho National Engineering and Environmental Laboratory (INEEL). Some residual contamination will remain in place and in the soils after remediation, possibly requiring some form of contaminant isolation for the duration of the health and environmental risk. Table 2 gives a cursory summary of solid wastes across the DOE complex. It is difficult to acquire site-specific information, but R.E. Gephart (Pacific Northwest National Laboratory, 2000, personal communication) provided some information concerning the Hanford Site. "The Hanford Site contains about 700,000 cubic meters of solid waste buried in 75 landfills—containing 6 million curies of radioactivity (decayed to 1998) and 70,000 tons (6.3 × 10^7 kilograms) of chemicals. Materials include 0.4 tons (400 kilograms) of plutonium and 650 tons (5.9 × 10^5 kilograms) of uranium. A small percentage (about 3 percent) was stored in above-ground facilities. Sixty percent of Hanford's solid waste was buried before 1970."

As another example, highly radioactive residues separated during the processing of very rich uranium ores from the former Belgian Congo (now Zaire) are presently stored at the Niagara Falls Storage Site in Lewiston, New York, buried under an interim cap to inhibit influx of moisture from precipitation and to decrease outflux of radon gas. A study by the National Research Council (1995a) recommended that these highly radioactive residues be removed, treated, and disposed off site to reduce the potential long-term risk to the public, rather than be covered with a "permanent" cap. No matter how well the cleanup is conducted at sites such as these, a certain amount of residual contamination will remain behind that may well require long-term monitoring and barrier maintenance.

TABLE 2 Summary of Solid Waste Across the DOE Complex (from *Linking Legacies* [U.S. Department of Energy, 1997b, pp. 39, 47, 53, 54])

Waste Category	Volume (m^3)	Curies
Low-Level Waste[a]	3.3×10^6	50×10^6
Low-Level Mixed Waste[b]	0.15×10^6	2.4×10^6
Transuranic Waste[c]	0.22×10^6	3.8×10^6
Total	3.67×10^6	56.2×10^6

[a] Low-level waste (LLW) includes all radioactive waste that is not classified as high-level waste, spent nuclear fuel, transuranic (TRU) waste, uranium and thorium mill tailings, or waste from processed ore. In volume, most low-level waste consists of large amounts of waste materials contaminated with small amounts of radionuclides, such as contaminated equipment (e.g., gloveboxes, ventilation ducts, shielding, and laboratory equipment), protective clothing, paper, rags, packing material, and solidified sludges. However, some low-level waste can be quite high in radioactivity.

[b] Low-level mixed waste (LLMW) contains both hazardous and low-level radioactive components. The hazardous components are subject to the Resource Conservation and Recovery Act of 1976, as amended (RCRA), whereas the radioactive components are subject to provisions in the Atomic Energy Act. LLMW results from a variety of activities, including the processing of nuclear materials used in nuclear weapons production and energy research and development activities.

[c] Transuranic (TRU) waste is waste containing more than 100 nanocuries of alpha-emitting transuranic isotopes per gram of waste, with half-lives greater than 20 years, except for (a) high-level waste, (b) waste that DOE has determined, with the concurrence of the EPA, does not need the degree of isolation required by 40 CFR 191, or (c) waste that the U.S. Nuclear Regulatory Commission (USNRC) has approved for disposal on a case-by-case basis in accordance with 10 CFR 61. TRU waste is generated during research, development, nuclear weapons production, and spent nuclear fuel reprocessing.

Soil

Contaminated soil is present at the major DOE weapons sites, especially at Hanford, Savannah River, Idaho Falls, Rocky Flats, the Nevada Test Site, and Oak Ridge, where large, diverse, and highly radioactive operations were carried out. There is also substantial soil contamination at sites such as Fernald, the gaseous diffusion plants, and the uranium mill tailings sites. Because surface soil contamination has a tendency to spread, it can increase the volume of the contaminated subsurface zones albeit at reduced concentrations of contaminants. Reduction of the contamination in soil may be achieved by chemical and/or physical means (National Research Council, 1999e). Radioactive contaminants in soil are generally removed to an on-site or remote burial ground; rarely is soil treated in situ. When the contaminants are organic compounds the soil may be decontaminated to regulatory limits by a number of means, such as "stripping" by passing a stream of air through it, thermal destruction by heating the soil batch-wise, or microbial action. In some cases, the soil volume and nature of contamination is such that selective leaching or other segregation of the contaminant may be feasible to reduce the soil contamination to the desired end state.

The Nevada Test Site (NTS) presents an example of soil contaminated by nuclear testing. The area contaminated is extensive, and it is not contained by any sort of engineered barrier. However, because decontamination of the site would be prohibitively costly using currently available technologies, at present no subsurface contamination reduction program is planned. Part of the site is used currently for disposal of low-level radioactive waste. DOE has not formally announced the end state for future land use at NTS other than to maintain a mission objective of possible resumption of weapons testing.

Currently there are no set standards for soil decontamination. The National Council on Radiation Protection and Measurements (NCRP) (1999) published screening limits for radionuclides in soil that relate an effective dose to a critical group to a corresponding soil contamination level. The screening levels are consistent with the NCRP recommendation that the maximally exposed individual should not exceed 0.25 mSv per year (25 mrem per year) from any single set of sources. Different screening levels are derived for various land uses from farming to commercial use. However, these limits are stated *not* to be used as cleanup standards on the grounds that they apply to the maximally exposed person and are conservative. A cleanup standard for plutonium in surface soil of 200 pCi/g (7400 Bq/kg) is in use as a de facto standard at NTS. This concentration is estimated to give an exposure of 100 mrem per year for a full time resident. The USNRC has promulgated cleanup standards for radioactive contamination in soil that are applicable to decommissioning of USNRC-licensed sites. The USNRC ground cleanup standard is based on individual radiation exposures of no more than 25 mrem/year to an average member of the critical group (10 CFR 20.1402). However, the EPA objects to this standard and recommends a limit of 15 mrem/year from all pathways, with no more than 4 mrem/year through the drinking water pathway for decommissioned sites. The appropriate contaminated soil remediation action is determined by the details of the particular situation, both with respect to the degree of health and environmental threats, the availability of practicable remediation technologies, and the financial resources to implement the technologies.

It should be noted that all of these potential standards for soil contamination are for calculated doses, derived by using various models to predict the radiation doses resulting from the contamination. These radiation doses are all very low when compared with typical background radiation doses and variations in background radiation, making the contamination doses extremely difficult to measure. Although the federal agencies involved (DOE, USNRC, EPA, and Department of Defense) have not agreed on standards for soil contamination, they have collaborated on guidance for radiological surveys conducted to demonstrate compliance with such a standard in the report Multi-Agency Radiation Survey and Site Investigation Manual (MARSSIM) (U.S. Department of Defense, U.S. Department of Energy, U.S. Environmental Protection Agency, and U.S. Nuclear Regulatory Commission, 1997).

Groundwater

Many soil contamination problems become water contamination problems through solubilization or suspension of the contaminant(s). The magnitude and severity of the water contamination problem is strongly influenced by the nature of the site, especially the composition and structure of the local geological formations and the climate. An essentially dry climate such as prevails at the Nevada Test Site, the Idaho National Engineering and Environmental Laboratory, and the Hanford Site poses very different problems from those regions having a wet climate such as found at the Oak Ridge Reservation and the Savannah River Site. It is also important to consider the rate at which the contaminants move and the likelihood of contaminated water being used for agriculture, by wildlife, or for domestic residential purposes. If these events are likely, the cleanup problem takes on a greater urgency. The Columbia River is important in this regard at the Hanford Site, as is the Savannah River at the Savannah River Site and the aquifer underlying INEEL. At other sites, such as the Fernald and Mound sites, a major aquifer is at risk.

Pump-and-treat systems, which involve installing wells at strategic locations to pump contaminated groundwater to the surface for treatment, are by far the most commonly used and proposed decontamination treatment for contaminated groundwater. Studies indicate, however, that pump-and-treat systems may be unable in most cases to remove enough contamination to restore groundwater to drinking water standards, or that removal may require a very long time—in some cases centuries (National Research Council, 1994c). In the cases where the contaminant is a relatively short-lived radionuclide (e.g., tritium), it is possible to conceive of a situation where pumping and treating the contaminated groundwater to storage or to recycle it repetitively might provide enough time for radioactive decay to reduce the contaminants to acceptable levels (for example, the "pump and reinject" system used at the Savannah River Site to deal with tritium). Radioactive decay in this case is a form of "natural attenuation"[2] (National Research Council, 2000a). Although pump-and-treat is apparently intended for use at

---

[2] Natural attenuation usually means that no action is taken to treat the contamination and that radioactive decay or natural destruction of an organic pollutant alone takes care of the problem. However, it may be interpreted more broadly to include natural flushing (and dilution) of contamination by the movement of water across the contaminated zone or object.

many of the DOE sites (see table in Appendix B), it may be effective for some purposes such as control of migration, but not for others such as complete removal of the contaminated fluids. In the many instances where long-term pump-and-treat methods are proposed, there is a need for equipment maintenance, monitoring, and all the operations involved in packaging the removed contaminant, transporting it to an acceptable disposal area, and disposing of it. In some cases, for example, organic contaminants, further treatment such as destruction by incineration might be required. Management of these operations will probably have to be carried out at a large number of sites, sometimes for decades or centuries. In addition, it should be borne in mind that in situ methods such as pump-and-treat are limited in the degree to which they can remove contaminants. In some cases the number of cycles necessary to reach the desired level of contamination will be impracticably large.

Nuclear Weapons Test Sites

Somewhat unusual in terms of contamination are sites where nuclear weapons have been tested underground, on the surface, or in the air. The committee visited the Nevada Test Site (NTS) at the request of DOE and because it is representative of a large DOE site where substantial amounts of radioactive and hazardous materials exist and are likely to remain. The NTS at once epitomizes the activities (e.g., ongoing operations, reindustrialization, cleanup, and the need for long-term stewardship) that are required at many of the DOE sites. Thus, a short description of the NTS situation as it relates to stewardship is given in Appendix F to give the reader a better appreciation of how the integrated set of activities at an actual site relate to the concepts presented in this report. The committee was unable to identify any specific commitment or process that would result in future re-examination of the major features of site remediation decisions being made today for the NTS, although decisions will be made on specific details (e.g., cleanup levels for specific locations) on a continuing basis. There appears to be little driving force for such reconsideration at present. Thus, the destiny of the NTS appears to be a limited number of remedial actions consistent with reindustrialization in selected portions of the site, followed by an indefinite period of institutional control in anticipation of possible future resumption of testing.

Surface Structures and Equipment

A very large number of contaminated structures exist on DOE sites, some of which are destined to be dismantled and disposed, while others are intended to be made available for use by industry. There are many firms devoted to decontaminating structures, and federal guidance for decontamination is evolving (U.S. Atomic Energy Commission, 1974; U.S. Nuclear Regulatory Commission, 1998 and in review; Federal Register Notice 63FR64132, 1998; U.S. Department of Energy, 1995b). However, the decontamination process may produce airborne particulates and/or liquid waste and contamination. In the abstract, the "best" decontamination treatment depends on the likely future use of the materials being decontaminated (National Research Council, 1998b), but future uses are often difficult or impossible to predict.

CONSTRAINTS AND LIMITATIONS

Technologies that will achieve a desired future state may be very expensive and produce unacceptably large volumes of secondary waste or their application may be necessary for impracticably long periods of time. In such cases it may not be practicable to achieve the desired level of decontamination. This puts greater requirements on contaminant isolation and stewardship activities. Therefore, a reasoned judgement is required of the technical, fiscal, safety, and regulatory aspects of the available technologies prior to deciding which one to deploy. The cost, risk, and systems analyses of such technologies should not proceed sequentially, but simultaneously, with strong interactions among them. It is important to recognize that although a risk assessment strives for an accurate and quantitative evaluation of risk, risk assessment is inherently a subjective process that is based on assumptions determined by the policy preferences of the assessor. Furthermore, the uncertainty associated with the resulting calculations may be very large due to a number of factors, including incomplete characterization of the site or contaminants, limitations on the ability to validate models of physical, chemical, and biological processes during contaminant transport and uptake, uncertainty and variability in the values of the parameters required by the models, and uncertainty in the quantitative estimates of health effects to human populations (Harley, 2000). Because of these factors there should also be consideration of contingency scenarios that would reflect different policy preferences and accommodate uncertainties in technologies, risks, and funding levels, as noted above. In addition to the above considerations, any specific decontamination technology would need to meet regulatory requirements, and may be subject to non-technical constraints.

Treatment of groundwater is a major concern because of the pervasiveness of groundwater contamination and the difficulty of effectively dealing with the contamination problem. Many of the major DOE sites have groundwater contamination problems that can affect rivers, lakes, and aquifers. Pump-and-treat cannot be relied on as a universally applicable technology for the indefinite future, and it is not clear what the follow-on treatments should be. The EPA and states have set groundwater concentrations of radionuclides for "safe" drinking. New standards are in the process of being promulgated for such elements as uranium and radon. As research on radiation carcinogenesis provides better quantitative data on the health effects of very low-level radiation exposures, risk guidelines may change (Jaworowski, 1999[3]). Similarly, successful decontamination of structures and materials is complicated by several limiting factors. The state of present technologies is the most pressing limitation, but physical structures and regulatory standards also present problems.

FUTURE DIRECTIONS FOR IMPROVEMENTS

It is likely that improvements in methods for characterization of contaminants will lead to changes in the selection and priorities of sites and facilities for cleanup. In this connection there is a need for new and improved methods of chemical speciation of contaminants. Cost reduction of characterization is highly desirable, as is increased sensitivity and speed. For example, new and improved methods for decontamination that reduce the amounts and risks of secondary wastes and reduce costs are needed, as are methods for rapidly, efficiently, and economically measuring the amounts of residual contaminants. Similarly, groundwater cleanup by methods other than pump-and-treat is highly desirable. Passive treatment systems and subsurface treatment walls show some promise for containment of some health and environmental threats.

The amount of contamination persisting at a site for long-term management (some for hundreds or even thousands of years) will be determined by the level of remediation that has been accomplished (based on such factors as budget, risk to the public and the environment, technical capability, regulations, and planned future land use) and the natural lifetime of remaining constituents (by such processes as natural attenuation, decomposition, biodegradation, or radioactive decay). As a consequence, decisions will be made at some time between the cost and risk of remediation and of long-term control and management. Such decisions will have to be revisited over time, based on new understanding of the contaminated environment and the new technology achieved from a continuing commitment to support of science and technology development directed toward environmental management and better understanding of the risk implications of social changes.

At most DOE facilities visited by the committee, a concern about the lack of appropriate and adequate data on the waste for modeling its migration into the environment was expressed. It is axiomatic that trustworthy decisions should be made based on sufficient, high-quality data. To solve the data deficiency, specifically for radioactive waste, requires qualified scientific and technical measurement groups and equipment that can assist in the characterization of the contamination across sites. Such a group, having expertise in radiation detection and measurement, could develop detailed knowledge of each site and identify similarities in types of contaminants and in physical properties of the environment. This would avoid duplication of effort among sites, provide a pathway for new generic instrument development and modeling, and augment sharing of novel equipment. Collaboration across sites produces an important gain in efficiency in site characterization measurements such as types of contaminants and environmental properties, as opposed to ad hoc individual site effort.

---

[3] This article elicited many comments in the Letters section of subsequent issues of *Physics Today* (e.g., April and May 2000).

The criteria for dealing with residues in liquid waste tanks should be amplified and refined, working with the USNRC, EPA, and the states. The changes should be based on consideration of the difficulties in characterizing the residues and their distribution. Consideration should be given to postponement of some tank closures to develop more effective residue characterization and extraction methods.

Different sites use different contractor laboratories or in-house measurement procedures for quality control. In order to have trust in any data collected, there should continue to be a long-term data comparison program among laboratories. In addition, experts are needed to undertake a quantitative and realistic evaluation of the potential health risks at each site, taking into account the natural background at that site. To accomplish this requires establishing a relationship with the regulatory organizations. Discussions, studies, and actions should take place for the purpose of reviewing existing compliance guidelines and determining any appropriate research necessary to quantify the risk of cancer and other health problems from low level exposures to be used to guide decontamination operations.

4

Contaminant Isolation

Contaminant isolation is the second of the three measures embedded in the long-term institutional management approach. This chapter addresses current technologies and methodologies used for isolation, desired characteristics of such measures, constraints and limitations for their application, and future directions for improvement, including the role of scientific and technical research.

Contaminant isolation refers to measures intended to prevent or limit contaminant migration into the environment adjoining a site. It becomes a necessary component of long-term institutional management in part because of the limitations on contaminant reduction discussed in the previous chapter. Contaminant isolation measures consist of engineered barriers, but also include groundwater pumping (hydraulic barriers) and waste stabilization approaches. As a group, these measures must be planned and coordinated closely with contaminant reduction measures, since the need for them is driven by the extent to which contaminant reduction measures are feasible or effective in reducing risk. Once in place, the ongoing effectiveness of contaminant isolation requires monitoring and maintenance and application of other aspects of the institutional management system. Over the longer term, monitoring of the groundwater and the unsaturated (or vadose) zone, as well as surface water, becomes important whenever contaminant isolation measures are in use.

DESCRIPTION OF THE TECHNOLOGIES

A recent report from the National Research Council (2000b) found increasing use and acceptance of waste containment and stabilization at U.S. Department of Energy (DOE) sites in recent years. Containment can be the low-risk, low-cost option of choice for some problems. Nevertheless, understanding of the long-term performance of containment and stabilization systems is limited, and there is a general absence of robust and cost-effective methods to validate that such systems are installed properly or that they can provide effective long-term protection.

Engineered Barriers

Engineered barriers, either on the surface or subsurface, are generally used to limit the contact of surface water or groundwater with wastes and migration into the surrounding environment. In special cases they may be used to limit the release of contaminated fluids and gases from leaking waste storage tanks, liquid waste transfer systems, or buried wastes. By far the most common engineered barrier is the surface barrier, often called a "cap," which is placed over waste deposits (see Sidebar 4-1). Surface barriers typically have multiple layers, with natural and synthetic materials of differing sizes and composition chosen to stabilize the barrier, prevent intrusion by animals and plants, limit movement of wastes, prevent infiltration of water into the waste deposit, and provide a mechanism to slow the release of radioactive or toxic gases. Vegetation is often planted to stabilize the top layer of the barrier, enhance evapotranspiration, and minimize water infiltration through the barrier; in some cases, however, vegetation may increase infiltration by slowing runoff. Two major wildfires in 2000 at the Los Alamos National Laboratory and the Hanford Site demonstrate the fallibility of barriers that depend on vegetation for stability; loss of vegetation from such fires may cause possible releases due to unpredicted environmental exposure and subsequent erosion. An example of a different approach is found with caps over uranium mill tailings burial sites, such as at Rifle, Colorado, that are sculptured to promote runoff of rainwater.

Subsurface barriers are not in widespread use at DOE sites, but they are receiving increasing attention as the problems of waste infiltration and transport by groundwater require more attention. Subsurface barriers may be either vertical or horizontal, and may function in several ways, depending on the specific nature of the groundwater transport situation. One approach is to divert water physically around or away from buried waste. Less common applications include chemical alteration or retention, attenuation, or destruction of wastes as they pass through a permeable barrier. Subsurface barriers may be either physical or chemical in their basic mechanism of waste retention. Mechanical barriers may include "walls" of concrete or metal, fused soil, or even horizontal barriers of concrete under such objects as leaking tanks. Chemical barriers formed from materials that react chemically with radionuclides or toxic materials may retain them or retard their movement by groundwater. Examples of relatively simple chemical barriers are clays such as bentonite and clinoptilolite that bind cesium ions and other ionic species of concern, and thus slow their movement. More complex examples might be phosphate-bearing materials to bind phosphate-insoluble ions chemically, or barrier materials capable of chemically reducing ions whose reduced ionic forms are much less mobile than the oxidized forms (e.g., technetium and neptunium).

Fusing the soil-containing contaminants into an impermeable or near-impermeable mass can protect the contaminants from intrusion and water transport. Usually the soil is composed predominantly of sand, and melting and fusion is accomplished through electrical resistance heating. A similar but temporary engineered barrier may be formed by the freezing of soil containing water. This type of barrier finds application where contaminated water is likely to leak from a container or other source during transfer of liquid. Subsurface barriers are sometimes made by injecting grout into the soil. The grout may incorporate materials such as zeolitic clays to bind certain mobile species. Alternatively, the subsurface barrier may be made of clay, without the use of grout as a host material, or it may include flexible synthetic membranes.

The use of engineered barriers for contaminant isolation was the subject of a recent workshop and report. During discussions at a joint National Research Council and DOE Workshop on Barrier Technologies for Environmental Management (National Research Council, 1997a), several recurring themes arose:

- The importance of proper installation techniques and quality control measures during construction, including the use of contractors with demonstrated experience and skill.
- The insufficient knowledge of effective lifetimes for barrier materials and systems.
- The importance of periodic inspection, maintenance, and monitoring, both short- and long-term, of containment barriers.
- The dearth of barrier performance monitoring data, and consequently the importance of compiling data on both successful and unsuccessful barrier installations.
- The advantages of using barriers in combination with pump-and-treat approaches to increase effectiveness.

Other good sources of information with respect to engineered barriers include Gee and Wing (1994), Rumer and Ryan (1995), Rumer and Mitchell (1996), and a recently published report on groundwater and soil cleanup issued by the National Research Council (1999e).

SIDEBAR 4-1

HANFORD SITE GROUNDWATER/VADOSE ZONE INTEGRATION PROJECT
(by Shlomo P. Neuman)

Since 1959, 67 of the Hanford Site's 149 single-shell high-level waste tanks have leaked or are suspected to have leaked about one million gallons (about 4 million liters) of waste into the ground. For years, the sorptive ability of sediments was expected to hold most leaked waste high above the water table. Upon closer examination of groundwater chemistry, contaminant distributions beneath tank farms, and geophysical data collected in wells, the U.S. Department of Energy (DOE) acknowledged that cesium-137, technetium-99, and cobalt-60 had migrated deeper than previously expected. Other tank-originated metals such as chromium, sodium, and nitrate are also likely in the groundwater. This admission resulted in negative national media attention and the reorganization of Hanford groundwater and vadose zone studies. It could also impact if and how waste is removed from high-level tanks for vitrifying and the technologies required to permanently close those tanks.

The Hanford Site Groundwater/Vadose Zone Integration Project (U.S. Department of Energy, 1998, 1999) was established in 1997 to coordinate and integrate the collection and interpretation of scientific information needed to deal with soil and groundwater contamination at the Hanford Site, Washington, on a site-wide basis. The project emphasizes characterization of the vadose zone, groundwater, and the Columbia River, and assessment of the risk that site contamination may pose to human health and the environment. Its intent is to help inform and influence key decisions by regulators and DOE concerning cleanup and environmental management of Hanford. To this end, the project aims to identify and address uncertainties and gaps in scientific understanding that influence such decisions, to initiate research that may help reduce these uncertainties and gaps, and to enhance the role of science and technology as a basis for site-related decisions. Through coordination and streamlining of site characterization efforts, the project hopes to eliminate redundancies and overlaps among these efforts. Additional project goals include development of risk assessment methods that are applicable across the site and the Columbia River system, rendering site information readily accessible to those who need it, facilitating public involvement in decisions concerning the cleanup and disposition of Hanford, and insuring independent technical reviews and management oversight of the Integration Project itself.

The Integration Project is envisioned as influencing Hanford Site decisions and operations such as high-level waste tank retrieval and closure, remediation of 200 Areas waste sites, and final closure of the Hanford Site, all toward protection of water resources, including the Columbia River. It is focused on five endeavors (U.S. Department of Energy, 1999):

- Integrate characterization and assessment work affecting long-term risk assessments.
- Assess the potential long-term effects of Hanford Site contaminants.
- Enhance the role of science and technology as a basis for cleanup decisions.
- Ensure productive involvement by parties interested in affecting Hanford's cleanup.
- Ensure independent technical reviews and management oversight of the Integration Project.

The contaminants at the reservation include radionuclides (e.g., carbon-14, chlorine-36, iodine-129, cesium-137, strontium-90, selenium-79, technectium-99, uranium-238, plutonium-239 and -240, tritium) and hazardous chemicals such as carbon tetrachloride, trichloroethylene, nitrate, nitrite, cyanide, and chromium. The sources of radioactive and hazardous waste contaminants to the vadose zone, groundwater, and, ultimately, the Columbia River, include planned disposal as well as leakage and spillage of high-level wastes from storage tanks and transfer lines in the central plateau (200 Areas) of the site. In addition, some tank liquids containing fission products from processing of the spent fuel for recovery of plutonium were directed into subsurface drainage "cribs," drains, ditches, and ponds that flowed directly into the soil

(continued)

> **SIDEBAR 4-1 (Continued)**
>
> (Gephart and Lundgren, 1998). Other sources of contaminants include burial grounds, injection wells, and ponds for holding the cooling fluids (water from the Columbia River) discharged from the reactors.
>
> If the Hanford Integration Project is allowed to go forward as envisioned by its organizers, it promises to improve the efficiency of site characterization and remediation efforts at Hanford and to enhance the role of innovative science and technology in the articulation and achievement of site cleanup and disposition goals. Major challenges faced by the project include (1) prioritizing its objectives to maintain and support research in the fundamental understanding and long-term remedies in addition to "applied science" and short-term technological fixes; (2) making efficient use of know-how and talent outside of Hanford, primarily at national laboratories and research universities, in the quest for improved science and engineering at the site; (3) developing and adhering to an ambitious but realistic schedule that balances short-term products with long-term needs; (4) ensuring sustained backing and financial support from DOE to operate effectively for as long as conditions at the Hanford Site require it; and (5) overcoming disincentives to scientific and technological innovation that existing contractual and institutional arrangements tend to foster.
>
> **REFERENCES**
>
> Gephart, R.E., and R.E. Lundgren. 1998 (September). Hanford Tank Cleanup: A Guide to Understanding the Technical Issues: Fourth Printing. Battelle Press, Richland, WA.
> U.S. Department of Energy. 1998 (December). Groundwater/Vadose Zone Integration Project Specification. Richland Operations Office DOE/RL-98-48, Draft C, Richland, WA.
> U.S. Department of Energy. 1999 (June). Groundwater/Vadose Zone Integration Project: Volume I-Summary Description; Volume II-Background Information and State of Knowledge; and Volume III-Science and Technology Summary Description. Richland Operations Office DOE/RL-98-48, Vol. I, II, and III, Rev. 0, Richland, WA.

Natural Barriers

In some cases the existing geology may be able to act as a barrier to migration. Wastes can be placed in such relatively impermeable strata as clay or some rock formations. For example, uranium ore residues have been interred in clay layers at the Niagara Falls Storage Site in New York (National Research Council, 1995a). In general, however, it is difficult to predict the long-term performance of such natural materials because of the general inhomogeneity in the formations (e.g., fractures, changes in the physical and chemical properties, inclusions in and intrusions into the formations) and changes over time that may result in the presence of difficult-to-detect preferred pathways for migration. The geological repositories for transuranic waste at the Waste Isolation Pilot Plant in Carlsbad, New Mexico, and the proposed geological repository for commercial spent nuclear fuel and high-level waste at Yucca Mountain, Nevada, both depend on engineered barriers to contain the waste in addition to the attributes of their natural geological barriers.

Groundwater Management and Hydraulic Barriers

Enhanced recharge and/or groundwater collection and extraction are often used to control the direction of local groundwater flow and to prevent the further migration of groundwater contaminants. Although not usually thought of as a barrier technology, groundwater collection and extraction, with subsequent treatment of the extracted groundwater (the pump-and-treat process), can in fact provide an effective but interim barrier to waste transport. The process provides hydraulic containment that prevents the further migration of radioactive and/or toxic materials in groundwater. When the amount of water to be treated is not large, and the waste material is not too dilute, the pump-and-treat process can provide a practical solution to many waste problems. The efficiency of the pump-and-treat process for restoration is greatly reduced when contaminants reside in (a) a heterogeneous medium with widely varying permeabilities and porosities, including fractures, (b) the unsaturated or vadose zone, or (c) a nonaqueous phase, especially dense nonaqueous phase liquids (DNAPL) (National Research Council, 1994c, 1999e).

Subsurface engineered barriers, such as slurry walls, are often used in conjunction with pump-and-treat systems to retard intrusion of uncontaminated water. Since pumping provides hydraulic containment, the barrier need not have a low permeability. The contribution of the slurry wall is to supplement hydraulic containment, thereby making containment easier and less costly to achieve.

The injection of water into a groundwater system to contain the migration of contaminants is another form of engineered barrier. In these cases, the injection of water is used to retard or change the local hydraulic gradient and in effect "contains" a contaminant plume. None of these technologies is particularly effective for managing the vadose zone, however (see Sidebar 4-1).

Waste and Contaminated Soil Stabilization

Stabilization approaches are often used to immobilize radionuclides and hazardous chemicals and thereby preclude their leaching and further migration from waste materials and contaminated soils. The approach is either to combine the waste with chemical additives such as lime, Portland cement, or fly ash to make a grout, or to provide electrical heating of the ground (e.g., in situ vitrification) to transform the waste or contaminated soil into a solid or glass from which radionuclides or hazardous chemicals (typically, metals are not easily leached) will not easily migrate (see Conner, 1990; Wilson and Clarke, 1994).

PERFORMANCE MONITORING OF ENGINEERED BARRIERS AND STABILIZED WASTES

Performance monitoring involves the continuous or periodic measurement of the effectiveness of the contaminant isolation system once it has been employed. The term "performance monitoring" often is associated only with the performance of a physical system (e.g., the reduced mobility of residual contaminants or the performance of technologies to isolate and/or clean up those contaminants). Monitoring is used to demonstrate the effectiveness of efforts to remove, treat, and contain contamination, but it is also used to support development of models of subsurface and contaminant behavior (National Research Council, 2000b). Approaches include groundwater measurement techniques available to conduct performance monitoring. These physical measurement techniques include groundwater monitoring (probably the most common), vadose zone monitoring, and cover and barrier monitoring (usually some form of vadose zone monitoring, but this can also include physical inspection). There is presently no well-established, reliable, and economic technology available to monitor effectively the vadose zone and heterogeneous media. This observation applies also to fractured subsurface media.

The Resource Conservation and Recovery Act of 1976, as amended (RCRA), the Comprehensive Environmental Response, Compensation and Liability Act of 1980, as amended (CERCLA), the Uranium Mill Tailings Remedial Action Program, and other waste disposal regulations require some type of post-remediation monitoring to be conducted at each site of residual contamination (see Appendix E). Monitoring requirements are usually spelled out in the site-specific documentation for each site. To date, it appears that relatively few site-specific post-remediation monitoring requirements have been defined at DOE sites. Numerous post-remediation issues remain to be resolved at many sites, including (1) What is to be monitored (e.g., soil, water, air)? (2) Where, how, and how often will monitoring be conducted? and (3) What conditions, if found, would necessitate further action (e.g., exceeding concentration limits or changes in hydraulic gradients)?

Monitoring serves two general purposes: to verify that the system being monitored is behaving as expected, and to compare monitoring results with pre-set limits for the purpose of standards verification. Monitoring will not be useful unless the results are examined critically, and conceptual modeling becomes a necessary adjunct to monitoring for establishing whether the system into which wastes have been emplaced is behaving as expected.

External Barriers

There are a number of reasons why monitoring and maintaining barriers, whether simple or complex, require effective institutional management. Near-surface disposal of chemical wastes and low-level radioactive waste usually requires the installation of a cap (i.e., a barrier above the waste to prevent the infiltration of water and subsequent transport of leached waste out of the disposal site, as well as to prevent human contact with the wastes). As previously noted, the cap may be composed of natural material such as clay, chosen for its resistance to infiltration, or man-made materials with greater expectations for performance. The cap may also be a multiple-barrier system. Whatever cap is chosen, its performance over time presents uncertainties. Settling can trap water, enabling greater infiltration through minor flaws in the cap. Undesirable vegetation may become established, with root systems that penetrate the cap. Careful attention to the cap and how wastes are emplaced can enhance confidence in continued good performance, but performance monitoring still will be needed. The monitoring systems associated with near-surface disposal of low-level radioactive waste are usually set up to monitor the water effluent from the burial cell. They should be designed to take into account uncertainty in the predicted performance of the barrier itself, such as through the use of systems to detect infiltration beneath the cap as supplements to monitoring intended to detect material leaking from the disposal cell.

Waste Stabilization

Another form of barrier for waste disposal is waste stabilization—that is, fixing the waste in a form that enhances its resistance to percolation and leaching. One example is the containment of low-level radioactive waste in grout or cement or vitrified into glass logs before disposal. The performance of the stabilizing medium as an additional leaching barrier is typically quite difficult to predict, however. Credit for regulatory compliance cannot be taken for the barrier unless performance can be demonstrated.

Another example is the use of multiple grout and cement barriers to fill an emptied high-level radioactive waste tank when its residue has been reduced to low-activity waste, a practice that has now begun at the Savannah River Site. The grout and cement barriers are intended to isolate the residue from water infiltration and leaching and to serve as a barrier to intrusion by animals, plants, and humans. Predicting performance in resisting water infiltration can be difficult because of uncertainties that include the degree to which the first layers of grout take up the residue, the water pathway effects of the cold joints between successive pours of grout, and the effects of preferential corrosion of the tank metal and penetrating structures (thereby offering a partial bypass path). Moreover, waste tank residue is likely to be highly radioactive and not taken up in the grout, so there is substantial uncertainty associated with the volumetric classification and average concentration of the waste and prediction of the isolation performance of the system. Finally, a key challenge to disposition of the tank residues in this manner is to obtain a determination of waste incidental to reprocessing to allow it to be classified and handled as transuranic or low-level waste (for example, the Savannah River Site) (U.S. Nuclear Regulatory Commission, 1999).

CHARACTERISTICS OF IDEAL CONTAMINANT ISOLATION MEASURES

Much has been written about the ability of engineered barriers and waste stabilization measures to last for the required time period necessary to manage the risk associated with leaving wastes in place. It is difficult to project limited performance data that exist much beyond a few hundred years, and even these time periods are very controversial. Ongoing maintenance will be necessary, including, perhaps, replacement of the system itself. In order to be effective, contaminant isolation measures should have the following characteristics:

- A design appropriate to the specific contaminant isolation requirements that provides the needed degree of protection and containment. The design should be developed with performance monitoring, maintenance, and repair needs in mind.
- A well-designed performance monitoring approach that addresses how criteria for failure are determined for the system selected and the specific site environment.

- A management and maintenance plan that specifies the types and frequencies of inspections and associated system repairs when needed, coupled with reasonable assurance that the plan will actually be carried out.
- Incorporation of adequate quality assurance and quality control measures during the planning and implementation stages. These measures are critical to success since the best-designed barrier can, and most probably will, fail if the installation of the system is compromised.

CONSTRAINTS AND LIMITATIONS

Engineered barriers and waste stabilization approaches, while potentially providing solutions to some of the most difficult waste management problems, are not without potential shortcomings. They can be expensive to build and install. Experience with some of the more novel applications of barriers is limited. The retrieval of high-level waste from storage tanks may leave significant residues of "incidental waste" in the tank. That waste may be low-level waste or transuranic waste in the DOE classification system. If that waste is to be fixed in place it must be provided with sufficient isolating barriers to assure adequate protection. If further, more aggressive waste extraction is not to be attempted, a custom-designed barrier system may be installed to stabilize the residue for in situ disposal, but doing so runs the risk of reducing future options for using later, better techniques. Current thinking for such waste stabilization at the Savannah River Site calls for initial injection of chemically reducing grout to fix the residue, followed by the addition of other cementitious materials to accessible areas inside the tank vault and the tank. The objective is to achieve a barrier system that is as robust as reasonably achievable, given the limitations of working with an existing tank system design.

Incomplete Understanding of Long-Term Performance

Perhaps the most important consideration in the use of engineered barriers and waste stabilization approaches in waste management is the fact that there is limited experience with most, if not all, of the systems being considered. The lack of experience with barriers proposed for use in some of the more demanding applications raises particular concerns. Concrete barriers can degrade with time, as can chemical barriers, which by their nature will be altered through chemical reaction with the wastes whose chemical nature they are intended to change. Barriers made of synthetic materials can also deteriorate over long periods of time; unfortunately, data on their performance over time are especially limited.

Need for Institutional Management

The limited lifetimes and effectiveness of barriers for the long term lead to the conclusion that institutional management will be required to ensure the effective performance of barriers, except when they survive long enough that natural processes, such as radioactive decay or, in the case of toxic organic materials, biodegradation, reduce risk to acceptable levels. Regardless of the design life of a barrier or stabilized waste, there is a need to confirm and ensure its effectiveness and durability over time. This assurance needs to be provided by the use of institutional management measures such as sampling and/or monitoring to determine if the barrier is, in fact, functioning as designed. As subsequent chapters will discuss, the effectiveness of institutional management measures should not be assumed; most if not all barriers will require the use of some kind of institutional management to ensure their efficacy and durability. Necessary institutional management measures include, for example, maintenance of monitoring stations and, in some instances, periodic inspections. Data need to be evaluated with reasonable frequency by individuals with appropriate expertise and appropriate corrective measures implemented.

FUTURE DIRECTIONS FOR IMPROVEMENT

The evolution in the sophistication of barrier design and technology over the past few decades has been remarkable. It is to be hoped that this progress will continue as the scientific and engineering community seeks to refine and improve engineered barrier materials and design approaches. Nevertheless, given the contrast between current contaminant isolation technologies and the characteristics of an ideal approach, there are several areas whose further research and development appear to have merit. New emphasis should be placed on the development of effective methods of performance monitoring. An adequate definition of "What constitutes failure?" should be determined, documented, and implemented. The procedure should also accommodate inevitable uncertainties that are inherent in the measurement process.

A corollary to the above is the realization that ongoing maintenance is critical to contaminant isolation effectiveness. Unfortunately, as will be discussed in later chapters, serious limitations in institutional management approaches may well mean that even careful planning will not be enough to ensure that maintenance will be carried out properly and that major failures will be remedied. Still, there is good reason to incorporate the need to accommodate repairs, including potential system replacement, into the initial design.

Finally, there will be an ongoing need for inspections, data collection and analysis, and decision making to determine whether the contaminant isolation technology is working or corrective action is needed. Monitoring approaches using trend analyses that accommodate uncertainties are emerging and their use is encouraged.

In summary, with few exceptions, contaminant isolation measures will be a necessary part of the overall remediation approach. Complete decontamination to avoid risk is rarely achievable. In some cases the contaminant may be moved into an isolation cell or impoundment designed specifically for the purpose (e.g., a uranium mill tailings impoundment). In other cases, the stabilization system design should be tailored to the existing configuration of the waste to be stabilized (e.g., waste residue in a buried high-level waste tank). In either case, the waste stabilization designer should take due account of the state of knowledge and uncertainties in waste isolation performance. Yet, good design may not be enough; it may be prudent to consider whether the waste should be *as isolated as reasonably achievable* (AIARA) (see Sidebar 4-2). In many cases, contaminant isolation measures cannot be relied upon to achieve their objectives without institutional management.

SIDEBAR 4-2

HOW CAN RADIATION EXPOSURES FROM WASTE DISPOSAL BE ALARA?
(by Robert M. Bernero)

For many years the international radiation protection community has advocated and followed three basic principles for protection against the potentially injurious effects of ionizing radiation, namely, (1) any practice that entails radiation exposures shall be justified, (2) strict limits shall be maintained for radiation exposures, and (3) radiation exposures shall be controlled to be *as-low-as-reasonably-achievable* (ALARA). Applying these principles in the control of nuclear activities results in setting a limit of 10 to 20 mSv/a (1-2 rem/yr) for radiation workers along with control practices that effectively keep these exposures ALARA, typically well below 5 mSv/a (500 mrem/yr). These radiation exposures are measurable.

For members of the public the limit is set at 1mSv/a (100 mrem/yr), a strict limit because the average member of the U.S. population is estimated to receive 3.6 mSv/a (360 mrem/yr) from background radiation. At these lower levels direct measurement is more difficult. Since public radiation exposure from nuclear facility operations is almost always due to effluents in the air or water pathway, controls to maintain public exposure ALARA are usually achieved by setting the point of compliance with the limit at a very close release point such as the exit of a ventilation stack. In this way there is substantial decrease in the actual radiation exposure at the site boundary. Reduction of the release source term is also available by filtration and other means to maintain releases and exposures ALARA.

In radioactive waste disposal a different situation is found. Decontamination or source removal can be conducted until measured residues have fallen to a level at which projected exposures of nearby populations are within protection limits and ALARA with respect to the cost or difficulty of further source reduction. For the deliberate disposal of large amounts of waste or for in situ disposal of wastes too difficult to remove, a system of barriers designed to inhibit waste migration from the site and consequent exposure of nearby populations is used. Typically, the barriers can be a system of packaging or stabilization of the waste itself to resist leaching, provision of caps or covers to divert water from coming into the waste, and liners or barriers to prevent contaminated water from leaving the site. Performance of such a system cannot be directly measured, but only predicted through performance assessments. Down-gradient monitoring cannot be relied upon to detect in a timely way a system failure resulting in a release of radionuclides. The selection of a good low-level waste disposal system creates a tension between demonstrating compliance with disposal requirements using the uncertain predictions of a performance assessment and the lingering need to adhere to the third principle of radiation protection, to maintain exposures, or releases, ALARA. The compliance limit for waste disposal is typically a fraction of the public health exposure limit, while only about a factor of 10 less is a negligible exposure, evidently ALARA. The uncertainties of performance assessment are too great to enable discernment of that factor of 10 for ALARA demonstration. In the early 1980s the new U.S. regulations for disposal of commercial low-level radioactive waste (LLW) were released. Explicitly related to near-surface disposal or shallow land burial of waste, they set compliance requirements for the emplacement of stable waste forms in a well-covered and well-drained site. Further enhancements are welcomed, but the sections of the regulations reserved for engineered enhancement, "other than near surface disposal" (10 CFR Part 61.51[b] and Part 61.52[b]) were never completed. Many expressed a desire for such features for U.S. sites east of the Rocky Mountains, where the waste was likely to be in the saturated zone near the water table.

At about the same time that the U.S. was adopting its LLW regulations, France was modifying its own methods for disposal of LLW. There, the new methods for engineered, near-surface disposal were developed and partially implemented at the Centre de La Manche, west of Cherbourg (now closed), and fully implemented from the beginning at the Centre de L'Aube, currently operating east of Paris. In the French system the regulatory requirements also call for stable waste forms in a well-covered and well-drained site. Compliance is measured by performance assessment conducted as if the LLW were buried in trenches. Since French sites are similar to many U.S. sites east of the Rocky Mountains, the uncertainties in performance assessment are similar.

However, in France the waste management authority has added many enhancements to the isolation capability of the site based on their judgements of cost effectiveness and good management. To begin, all waste shipments are tracked, with their assay, using bar code markers to their records of generation, transport, and disposal. At the disposal site many of the waste packages are tightly compressed to fit together in concrete cylinders sealed with bitumen. A new combined bar code is assigned to this package. The waste packages are typically packed closely into concrete near surface vaults, with grout added to fill the interstitial space. The excavation and array of waste emplacements are covered by a very large canopy building mounted on rails, so that the waste emplacement area is kept dry during operations. After the waste is emplaced and the vaults closed, it is covered with soil, a synthetic membrane, and a clay layer to keep out moisture, and topped with soil cover and vegetation to retain stability. The emplaced waste has an underlying leachate collection system as a precaution, to be monitored for 300 years, although one would not expect any leachate from the array. Thus, the French system does not rely on subtle differences in uncertain performance assessments to determine whether public exposures are ALARA. Rather, it implicitly uses a different but equivalent principle of protection. The systems attempt to render the waste *as-isolated-as-reasonably-achievable* (AIARA). It is difficult to quantify the effects of these enhanced barriers to waste movement, but they are especially valuable in enhancement of intrusion barriers and in isolation of dominant waste isotopes with half-lives of about 30 years (e.g., cesium-137 and strontium-90). The enhancements compensate for uncertainties in compliance performance assessment. This AIARA concept also can be applied to the disposal of high-level waste (HLW) deep underground. Actually the concept is implicit in the waste disposal statutes where near-surface disposal is authorized for LLW, but deep geologic disposal is required for HLW, providing a much greater degree of isolation.

Others have shown interest in the French approach for LLW isolation. In Spain, the El Cabril LLW disposal site is a near copy of the French Centre de L'Aube even though El Cabril is in a much drier, high elevation setting in southern Spain. In the U.S. no new "full service" LLW sites have been licensed since the Low-Level Radioactive Waste Policy Amendments Act of 1985, but most of the states east of the Rocky Mountains have indicated a desire to incorporate the engineered features of the French system, perhaps to render their waste AIARA. Nevertheless, the fallibility of bar code markers for maintaining records and identifying waste packages over periods of hundreds of years is recognized. In addition, the stability of governments over a period of 300 years is questionable. For example, a quick glance at the last 300 years of French history reveals a monarchy, five republics, a couple of empires, a war zone, and a hostile occupation. At each change in government, the survival of monitoring systems would be in extreme jeopardy.

> ## SIDEBAR 4-3
>
> ## THE HANFORD BARRIER
> ## (by James H. Clarke)
>
> At the Hanford Site in Washington, the Hanford Barrier (200-BP-1 Operable Unit) is a multilayer surface barrier constructed to evaluate design, construction, and performance features for use as in-place disposal of Hanford wastes in a semiarid to subhumid climate (Gee et al., 1994). The barrier features engineered layers designed not only to minimize the potential for water infiltration, but also to minimize the likelihood of plant, animal, and human intrusion, limit the release of vapors and runoff, and minimize erosion-related concerns. The design is intended by DOE to meet or exceed Resource Conservation and Recovery Act of 1976, as amended (RCRA), cover performance requirements, and to isolate wastes with transuranic constituents for a minimum of 1,000 years while being maintenance free. The approximate unit cost excluding testing and monitoring tasks was $320/m^2.
>
> Construction of a prototype began in the 200 Area of the Hanford Site in late 1993 and was completed in 1994. The prototype is highly instrumented and its performance under conditions of enhanced rainfall (three times normal precipitation conditions) is being monitored. The prototype design also incorporates two different degrees of side slope to provide information on the impact of side slope design on the competing objectives of infiltration prevention and erosion control.
>
> The Hanford Barrier prototype is an example of the kind of research and development so critical to the U.S. Department of Energy (DOE) environmental restoration mission. The data that are emerging from this program, along with data from DOE-sponsored studies at the Hill Air Force Base and at other DOE facilities, are essential to effective remediation decision making, not only within the DOE complex, but also at essentially all locations where engineered barriers are being considered for contaminant isolation. A four-year treatability test of a prototype of the Hanford Barrier (U.S. Department of Energy, 1999) that established a performance baseline of the barrier and its components has been recently issued.
>
> ## REFERENCES
>
> Gee, G.W., H.D. Freeman, W.H. Walters, Jr., M.W. Ligotke, M.D. Campbell, A.L. Ward, S.O. Link, S.K. Smith, B.G. Gilmore, and R.A. Romine. 1994 (December). Hanford Prototype Surface Barrier Status Report: FY 1994. Pacific Northwest Laboratory PNL-10275, Richland, Wash.
>
> U.S. Department of Energy. 1999 (August). 200-BP-1 Prototype Barrier Treatability Test Report. DOE/RL-99-11, Richland, Wash.

While there is typically a tacit recognition that engineered barriers and waste stabilization approaches have limited periods of effectiveness, these technologies are often employed with inadequate understanding of, or attention to, the factors that are critical to their success. These include the need for well-conceived plans for performance monitoring that identify and correct potential failures and plans for maintenance and repair, including possible total system replacement.

The committee stresses the importance of building into planning and design approaches a recognition of, and allowance for, uncertainty and fallibility. Contaminant isolation systems should incorporate an effective means of performance monitoring as close to the waste and contaminated soils as possible without compromising system integrity. Monitoring within the containment system will provide useful information to vadose zone and groundwater monitoring and improve the ability to provide an early warning of potential failure.

It is now widely recognized that the subsurface is a complex, multiscale, spatially variable natural environment that can never be fully characterized. Hence the results of even the most thorough site characterization and monitoring efforts are ambiguous and uncertain. The DOE should continue to sponsor targeted applied research efforts that not only address the critical knowledge gaps concerning contaminant isolation system design effectiveness and lifetimes, but that also incorporate improved ways of detecting and remedying potential failures. Specific examples of such research are found in two recent reports by the National Research Council (1999e, 2000b). With respect to performance monitoring there is a need for good data appropriate to the task, better and more affordable data collection technology to collect such data, and scientific research directed toward improved understanding of the factors that affect system performance.

The Hanford Barrier prototype (see Sidebar 4-3) represents an example of the difficulty of conducting a continuing program of relevant research projects until the real benefits (in this case, better understanding of performance over time) have been achieved. After its construction, the prototype barrier was monitored for only a few years before funding was terminated. Recently, some limited funding has been made available. While, the DOE should be commended for initiating the Hanford Barrier prototype research (and other similar efforts, such as research appropriate to caps for waste sites in humid eastern U.S. environments), a long-term commitment to research funding and priorities is needed to ensure that the resulting data are sufficient for projections of future performance and potential design improvements.

5

Stewardship Activities

Stewardship activities comprise the third of the three sets of measures used in long-term institutional management. This chapter addresses, first, the activities that encompass stewardship, then the constraints and limitations for its application, the characteristics necessary for a viable stewardship system, and, finally, future directions for improving stewardship, including research and development needs.

Stewardship in the broadest sense includes all of the activities that will be required to manage the potentially harmful contamination left on site after cessation of remediation efforts. Some stewardship activities have been considered in Chapter 4, specifically measures to maintain contaminant isolation and measures to monitor the migration and attenuation or evolution of residual contaminants. Other stewardship activities that will be considered in this chapter include:

- *institutional controls* (generally, use and access restrictions);
- *conducting oversight* and, if necessary, *enforcement*;
- *gathering, storing, and retrieving information* about residual contaminants and conditions on site, as well as about changing off-site conditions that may affect or be affected by residual contaminants;
- *disseminating information* about the site and related use restrictions;
- *periodically reevaluating* how well the total protective system is working;
- *evaluating new technological options* to reduce or eliminate residual contaminants or to monitor and prevent migration of isolated contaminants; and
- *supporting research and development* aimed at improving basic understanding of both the physical and sociopolitical character of site environments and the fate, transport, and effects of residual site contaminants.

Ideally, most of these activities would begin when contamination of a site, either purposefully or accidentally, is first identified. Consequently, many of the activities would have developed to some degree of maturity prior to the time that remediation of the site is determined to be complete.

Stewardship activities entail ongoing, periodic if not continuous, actions by people. These people may be representatives of federal, state, or local governmental agencies, Native American groups, or private businesses and other non-governmental organizations; they also may be individual landowners, tenants, neighbors, or other concerned private citizens. Issues with stewardship include not only *what* will be done, but *how* and *when* it will be accomplished, and *by whom*. For this reason, in the following two chapters (Chapters 6 and 7) we delve more deeply into contextual factors and institutional as well as technical capabilities and limitations. As Chapter 7 will note, while activities are the most visible component of stewardship, they rest upon legal, financial, and organizational structures and social and political factors that must work well for these activities to be conducted as expected. In the present chapter, we focus on the stewardship activities themselves.

COMPONENTS OF A COMPREHENSIVE STEWARDSHIP PROGRAM

An adequate, comprehensive stewardship program for a residually contaminated site (including land, groundwater, surface water, and facilities) should include most, if not all, of the activities listed above. These activities should be conducted for as long as the residual contaminants remain potentially hazardous. Each activity is described below in general terms that set forth what ideally should happen with long-term stewardship. Short-term activities (e.g., supplying bottled water when well water is contaminated) are not discussed here, because they would ordinarily be used only as interim safeguards, not as part of an extended plan for managing a residually contaminated site.

Institutional Controls

The subset of stewardship measures known as institutional controls consists mainly of restrictions on land use. They also include the legal means to obtain access to a site that has been transferred from the U.S. Department of Energy (DOE) to another entity for monitoring and follow-up remediation (e.g., through the affirmative easements mentioned in the next section of this chapter). However, institutional controls commonly are equated with use restrictions, perhaps because use restrictions can figure so importantly in the beneficial reuse of residually contaminated sites.

Use restrictions—including legal restrictions imposed through easements, covenants, zoning, or permit requirements as well as physical restrictions such as fences, signs, and guards—are a basic component of stewardship. They should be in place in every case where a site is not considered safe for unrestricted use, or where contamination has migrated off site, affecting resources such as water and soil and necessitating restrictions on their use. The extensiveness and intensity of implementation of the use restrictions should directly correlate with the severity of the risk to potential users of the site. In other words, use restrictions should meet agreed-upon objective criteria of what is needed to reduce the prospective risk to human health, safety, and the environment to an acceptable level.

The site's condition should be carefully considered in selecting use restrictions. The rationale for not removing contaminants from a site, thus requiring use restrictions, should be clearly demonstrated. The use restrictions should be specified in detail, with the input of both the people having authority to implement and enforce them and the people (e.g., the surrounding community) most likely to be inadvertently exposed or to otherwise have an interest in how the site is used. However, because some residually contaminated sites will remain hazardous far into the future, the impossibility of involving all such people also should be recognized.

As discussed later in this chapter, even the most carefully crafted use restrictions should not be relied upon to remain in effect over time. For these reasons, remedies that rely on use restrictions should include "reopener clauses" triggered by, for example, unanticipated and unacceptable changes over time in use or other feedback from a monitoring program or members of the public, improvements in technological capability to fully remediate the site, or changes in regulatory issues concerning exposure of hazardous and radioactive substances to the public and the environment. In addition, as suggested later in this report, efforts should be made to assure that external groups and interested citizens retain rights of oversight and influence over the organization or organizations bearing primary responsibility for the site. Because of their importance to stewardship, institutional controls are described more fully later in this chapter.

Conducting Oversight and Enforcement

Oversight and enforcement are safeguards to ensure that the stewardship activities are carried out effectively and in a timely fashion. Oversight and enforcement should be conducted by an entity or entities with the power to ensure that these activities are in fact being accomplished, and to impose sanctions or otherwise rectify the situation if they are not. Oversight and enforcement should be tailored to the specifics of the site. The frequency and extensiveness of scrutiny should depend upon the risk associated with failure to conduct one or more of the above activities effectively. Oversight and enforcement mechanisms also should take into account the history and nature of relevant governing authorities, using—and upgrading as necessary—public or private institutions with good track records of responsible stewardship. Oversight and enforcement should factor in complex local land ownership histories and patterns. Nevertheless, in each case oversight and enforcement should be conducted dispassionately and consistently, and mechanisms should be fully integrated into federal, tribal, state, local, and private regulatory systems.

Gathering, Storing, and Retrieving Information

Information management includes the gathering, storing, and, importantly, *retrieving* of relevant site information when it is needed. At a minimum, the information to be managed will need to be informative about the nature, extent, and duration of risks from residual contamination (including hazard characterization or suspected health effects and exposure pathways), contaminant reduction and isolation efforts on the site, monitoring data associated with these efforts, use restrictions in place, and information about the entities responsible for implementing, overseeing, enforcing, and modifying the site's long-term management plan. As discussed further below, experience shows that paper records are easily lost in archives, and electronic data storage media are now changing over periods even shorter than a decade (ICF Kaiser, 1998; Tangley, 1998).

Disseminating Information

Successful dissemination of the information discussed above is a necessary condition for oversight and enforcement, as well as for other purposes. Information should be directed to the people and organizations who have a need to know because (1) they are responsible for implementing or enforcing the site's institutional management plan, (2) they could be harmed by failures of the plan, or (3) they are part of a larger community with an interest in the plan's success. People in the first category would include federal, tribal, state, or local officials or private companies with legal responsibilities for the plan: They need to know what they are protecting, how long it must be protected, and for what reasons. People in the second category would include site users as well as others such as well drillers, farmers, or hunters who might need to be informed of use restrictions. People in the last category might include, for example, members of the medical community needing to know the extent to which local people are drinking contaminated groundwater or eating contaminated fish. The last category might also include concerned individuals or organizations that unofficially monitor the site to ensure that use restrictions are observed and that the site's management plan is being properly implemented. The information conveyed, the methods of conveying it, and the targeted audiences may need to change through time to continue to be effective. In a recent book, Benford (1999) discusses several projects studied to convey information across "deep time," meaning time scales of at least centuries. He points out that such efforts to plan ahead for centuries and millennia present challenges that simply cannot be met with present capabilities, and that solutions to such challenges would require a profound cultural shift.

Periodic Reevaluation of the Site Protective System

The total protective system in place—the activities just discussed, as well as the contamination reduction and isolation efforts discussed in the preceding chapters—should be comprehensively evaluated on a periodic basis, to determine how well they are working as a system. This periodic required reevaluation can not be assumed to occur and be effective because the past record of similar such reevaluations shows numerous deficiencies. Moreover, the present capability for assuring future performance is limited (Freudenburg, 1992; LaPorte and Keller, 1996). Still, the importance of periodic reevaluation is undeniable. If parts of the system are weak or have failed, then the efficacy of the system as a whole may be seriously compromised. In contrast, if all parts of the system are working well, or if it is clear that some system components are no longer as essential as they once were (e.g., given a reduction in the toxicity of the residual contaminants), then the people responsible for long-term management of the site can be reasonably confident that the system will be protective until the next comprehensive re-evaluation.

Legal requirements (e.g., the five-year review requirement under the Comprehensive Environmental Response, Compensation, and Liability Act of 1980, as amended—CERCLA) will set a minimum standard for the periodicity of such comprehensive reevaluations. However, these requirements are made with, at best, a rudimentary understanding of the nature and scope of each individual problem. More or less frequent comprehensive reevaluations may be appropriate for some sites; thus, this topic should be discussed and agreed upon by DOE, its regulators, the surrounding community, and those responsible for institutional management when its measures are put in place. Moreover, the frequency of comprehensive reevaluations should, over time, depend upon how well the system is performing; if it is performing poorly, more frequent reevaluations may be necessary.

In conducting a comprehensive reevaluation of the protective system, off-site factors (e.g., changes in surrounding land uses or hydrologic conditions) as well as on-site factors should be taken into account. In other words, the site's protective system should not be treated as an isolated phenomenon; instead, it should be thought of as part of a larger fabric that inevitably will experience physical and social changes over time.

Cultivating New Remediation Options and Developing Better Understanding of Site Contaminant Behavior

As has already been noted in this report, stewardship activities at many DOE sites have limited likelihood of remaining effective for as long as residual contaminants will remain hazardous. For this reason, the set of site stewardship measures is taken in this report to include support for both basic science and applied science research and development (R&D). It also includes the monitoring of scientific and technical breakthroughs in arenas beyond those controlled by DOE for their applicability to DOE sites. Including such elements among stewardship measures is intended to address beforehand the possibility that a site's protective measures might fail. It does so both by reducing the consequences of such failures should they occur and by reducing the probability of their occurring in the first place.

Over the long run, both monitoring newly emerging technologies for their potential application to sites with residual contamination and directly funding research and development on new technologies will serve to help reduce risk, thereby boosting the effectiveness of site remediation. More effective site remediation thus becomes a necessary condition for more effective site stewardship. Data collection for purposes of monitoring and site surveillance, as well as in support of site and waste characterization, is another set of activities that belong in this same category. Because these activities support improved understanding of site environments and the sources, fates, and effects of contaminants that remain on site, they also contribute to more effective site stewardship.

More generally still, the same can be said for support of basic scientific research aimed at improved understanding of sites and the fate and transport of residual contaminants on them. Adequate support for this last component of a comprehensive stewardship program is not likely to emerge on a site-specific basis. System-wide attention is necessary if basic limits in scientific understanding with the potential to undermine the effectiveness of institutional management programs at individual sites are to be addressed effectively.

A broad, nation-wide stewardship program must provide support for scientific research, both for the physical and social sciences, and for technology development that is directed toward reducing the risk to the public and the environment posed by residually contaminated sites. Such research and development should be conducted in conjunction with the remediation program prior to site closure, but should continue as part of long-term institutional management following closure. The overall goal of such research and development should be improved understanding, methodologies, and technologies that have the potential to reduce both the cost and risk. A number of recent reports by the National Research Council (e.g., 1994c; 1996e; 1997b; 1998b; 1999c,e; 2000a,b) provide some details of the research and development currently needed by DOE to accomplish remediation (both contaminant reduction and isolation) of its legacy waste sites.

TYPICAL INSTITUTIONAL CONTROLS

Although stewardship of residually contaminated sites in both the public and private sectors is now receiving considerable attention (see Appendix D), it was largely ignored until recently. To the extent that it did attract attention, institutional controls, particularly use and access restrictions, typically were the exclusive focus. As noted above, institutional controls are only one component of a total system of effective stewardship. Nevertheless, they merit special attention because of their importance. Under ideal conditions they can become the institutional counterpart to engineered barriers in preventing undue exposure to residual contaminants. For this reason, we provide a brief summary of typical institutional controls here. Much has been written on institutional controls; some of the more recent reports are described in Appendix D. In addition to these resources, DOE and EPA are currently preparing guidance documents on selection of institutional controls and their application.

Easements

Easements are based in property law. They are legal devices through which limits are placed on the use of property or through which people other than the property owner are allowed use of the property for a specified purpose. For easements to be enforceable, the property owner must grant a property right to another party, who then becomes the easement holder. This right is recorded with the appropriate governmental unit, to give notice to members of the public and any future purchasers of the property.

As suggested in the foregoing paragraph, easements can be affirmative or negative. *Affirmative* easements give the easement holder usage of or access to property owned by someone else. In the context of environmental remediation, an affirmative easement would allow the easement holder to come onto property owned by someone else to perform monitoring or a response action. For example, according to CERCLA Section 120(h)(3), when previously contaminated property is transferred by the federal government to anyone except a party potentially responsible for the contamination, the deed must contain a covenant warranting that any additional remedial action found necessary shall be conducted by the United States. An affirmative easement would likely be required to fulfill the terms of the covenant. *Negative* easements allow the easement holder to limit the owner's use of the property. A negative easement would allow the easement holder to preclude the property owner from activities such as well drilling, use of certain chemicals, or excavation below a certain depth in order to protect barriers, pump-and-treat systems, or other remediation activities taking place. This type of easement is more commonly called a deed restriction.

Conservation easements, a form of negative easement, allow the easement holder to dictate that the property can only be used for conservation-related purposes; they are recognized by almost every state (Korngold, 1984). Conservation easements overcome problems associated with traditional easements, and therefore may be useful at some contaminated sites (U.S. Environmental Protection Agency, 1998b). In some cases, a conservation easement can be enforced, not only by the easement holder, but also by others. Like other easements, however, conservation easements do not necessarily offer "perpetual" protection; they are subject to political influence (Ohm, 2000). Covenants are similar to easements in that they also require a conveyance of the interest in land. Covenants and easements differ in the formal requirements needed to effectuate them. Equitable servitudes have some similarities to covenants, and courts may "create" an equitable servitude if some of the formal requirements of a covenant are lacking.

Deed Notifications

Deed notifications are descriptions included in the deed to put future buyers on notice about some particular feature of the property. For example, deed notifications are required under CERCLA Section 120(h)(1) for any transfer of federal property if hazardous substances were known to have been disposed or released, or stored for a year or more on the property. They are required under the Resource Conservation and Recovery Act of 1976, as amended (RCRA), to inform future buyers that the property was used to manage or store hazardous wastes (see Appendix E). Deed notifications do not create enforceable use restrictions because they do not involve granting a property right; however, they often are considered an institutional control because they may serve to deter inappropriate uses. Deed notices, as well as easements and covenants, are subject to the vagaries of each county's recording system.

Zoning

Zoning is a local governmental authority (under the government's "police power") to designate and regulate land uses. In the context of remediation, zoning operates as an institutional control if it serves to restrict site uses to those that are compatible with the cleanup level that has been achieved. Traditionally, comprehensive land use zoning authority has been enabled by states and exercised by local governments; however, not all local governments have enacted zoning ordinances, and zoning decisions can be changed relatively readily by governing bodies.

Permit Programs

Permits or licenses can be granted by the appropriate local, state, or federal government entity to allow certain land use activities such as well drilling, excavation, blasting, mining, and construction. Permit programs function as institutional controls when they are relied upon to ban or restrict activities that could conflict with an approved site use. Reliance on permits to serve as institutional controls would be based on the authority and capability of the permit program to implement and enforce the program effectively. The permitting entity would need to have sufficient information to know why, where, and for how long a permit should be granted.

Fences

Fences are fixed structures that serve as boundaries or barriers. Ideally, the degree of impenetrability of the fence (e.g., a three-strand wire fence versus an eight-foot cyclone fence ringed with razor wire) should be commensurate with trespasser interest—at least to the extent to which we can know this in advance—and with the harm that could ensue if the fence were breached. Fences that are virtually impenetrable by humans, however, will not necessarily stop other species and thus may not prevent the migration of terrestrial animals or plants onto or off of a contaminated site.

Signs

Signs as institutional controls consist of the message and the material used to convey the message. Sometimes they must last a very long time. For example, at the Waste Isolation Pilot Project (WIPP, a deep geologic repository for transuranic wastes in New Mexico) DOE intends to use both records and physical markers to warn future societies about the location and contents of WIPP in order to help deter inadvertent intrusion over the coming millennia (U.S. Department of Energy, 1999), although independent observers have evaluated the feasibility of this effort with skepticism (Erikson, 1994). To be effective, a sign's message would need to be understandable by all intended audiences for the length of time it must convey information, and the sign material would need to endure for that same time period or to be properly maintained. The sign's message and material would need to be periodically evaluated for its effectiveness and durability and modified as warranted.

Government Ownership

The federal, tribal, state, or local governments that own contaminated sites can use their ownership rights to exclude all external use of the site or to impose use restrictions through leases or contracts. All government ownerships are not equal, however. Often, surface rights are split from mineral rights or water rights. At the federal level, some land is classified as "public" while other land is classified as "acquired lands." State land ownership is similarly complex. For example, U.S. western states upon being admitted into the union were given certain lands for schools that require management by state land boards to maximize income. The use of these lands is limited by state constitutions. Other lands acquired by state governments, such as office buildings, are simply owned by the state in the same fashion as land is owned by a private party. Land ownership complexities must always be carefully analyzed, both at sites themselves and as they affect adjacent land that might serve as buffers.

Leases

A lease can serve as an institutional control by requiring parties to observe use restrictions and other conditions. Sites owned by the federal government may be leased to other public or private parties, with the lease terms stipulating such things as water use restrictions, approved access routes, and construction limitations. Violations of the lease terms then need to be addressed in the courts.

CONSTRAINTS AND LIMITATIONS

The above general descriptions of the components of a comprehensive stewardship program, as well as the somewhat more detailed descriptions of typical institutional controls, are meant to give a sense of the range of activities that could and often should take place in conducting long-term stewardship of a residually contaminated site. Nevertheless, the efficacy of these activities is by no means assured. Problems that can arise in conducting these activities are noted below. Underlying contextual and structural factors contributing to these problems are discussed in Chapters 6 and 7.

Institutional Controls

Because efforts to maintain land use restrictions on private lands run the risk of coming into conflict with property rights, their long-term viability remains questionable. Private property rights are strongly supported in the U.S. Constitution, and the 'takings' issue is frequently raised when the use of private property is restricted in the name of protecting broader public values. The appropriate balance to be struck between protecting private property rights and the exercise of police powers in the name of public health and safety protection is a major area of inquiry in constitutional law and an area in which the courts have been especially active. The viability over time of land use restrictions is likely to be especially questionable in cases where contamination levels are not high enough to prohibit all public access but not low enough to permit unrestricted use. Often the real issue is not *whether* use restrictions will eventually fail, but when and what the *consequences* will be when they do.

Currently, ways to strengthen institutional controls for residually contaminated sites are being explored (English et al., 1997) (see below). While these improvements would make institutional controls more robust, past failures are worth noting. A few examples follow:

- In 1953, Love Canal was transferred from Hooker Chemical to the Niagara Falls School Board. The board gave assurances that no construction would take place in landfilled areas, and a deed notice was placed in the land records. Despite these measures, however, adjacent land was developed for housing soon thereafter, with homebuyers later reporting they had never been informed of the hazards or the deed notice. Within just two years, an elementary school had been erected and opened on top of the former Hooker Chemical landfill (Gibbs, 1982; Hersch et al., 1997; Levine, 1982; Mazur, 1998) (see Sidebar 5-1).
- In Oregon, houses were built on a closed landfill, even though the state had previously notified the county that the site should not be used without state approval. After the problem was discovered by state employees, residents' wells were sampled and found to be contaminated (Pendergrass, 1996).
- At the DOE Oak Ridge Reservation in Tennessee, land sold by the federal government in the early 1990s was to be used as a golf course. The deed prohibited use of groundwater that came from the Y-12 plant and was contaminated with organic chemicals (trichloroethylene). Within just a few years, however, DOE discovered that a well was being drilled to irrigate the golf course. Fortunately, DOE discovered this problem and, since that time, has upgraded its oversight regarding deeds of surrounding property.

SIDEBAR 5-1

LOVE CANAL, NEW YORK: AN EXAMPLE OF FAILED STEWARDSHIP
(by William R. Freudenburg)

The Love Canal was originally proposed by William T. Love in 1892, intended to harness the water of the upper Niagara River into a navigable channel—canal—then thought to be the future of industrial transportation, plus a 280-foot waterfall that could be used to generate cheap hydropower (Gibbs, 1982; Levine, 1982; Mazur, 1998). The canal would have been six to seven miles long, but it was abandoned after only about half a mile had been dug. In 1920, the land was sold at public auction and, after serving as a swimming hole and ice-skating rink for the few people who lived nearby at the time, it ultimately became a municipal and chemical company disposal site. The Hooker Chemical Company, which is generally seen as having been the major user of the site, began purchase arrangements for the site in 1941, started using it for dumping of chemical and hazardous waste from its manufacturing operations in 1942, and completed purchase in 1947. Hooker (which was the largest industrial enterprise in Niagara Falls in the 1970s, employing some 2,400 people and ultimately becoming part of Occidental Chemical) acknowledges having dumped some 20,000 to 25,000 tons of chemical wastes into the Canal. The city of Niagara Falls also used the site for dumping municipal wastes, and residents report that the U.S. Army used the site for dumping as well, although the Army denies having done so. Over 200 compounds had been identified in the Canal by the early 1980s; the largest component (roughly 25 percent by weight) was benzene hexachloride, a waste product from producing the insecticide lindane. This and many of the other compounds in the Canal were recognized by Hooker as having been toxic.

In 1953, after filling the Canal, Hooker covered the filled canal with dirt, selling it to the local Board of Education for $1.00. According to most accounts, the deed contained stipulations that warned of potential hazards and announced that if anyone was injured by the wastes, Hooker would not be responsible. Did it take hundreds of years for these stewardship measures to fail? Hardly. Home building began off-site (i.e., adjacent to the 16-acre rectangle that had once been the Canal) quite soon after the land changed owners. By 1955, only two years after the transfer, an elementary school had been constructed and opened on top of the hazardous chemical dumpsite. By the late 1950s, residents had begun to complain about sickness, odors, black sludge, and symptoms such as chemical burns on their children. It took nearly two decades before their complaints were taken seriously by the relevant governmental and health officials.

REFERENCES

Gibbs, L. M. 1982. Love Canal: My Story. State University of New York Press, Albany, N.Y.
Levine, A. 1982. Love Canal: Science, Politics, and People. Lexington, Books: Lexington, Mass..
Mazur, A. 1998. A Hazardous Inquiry: The Rashomon Effect at Love Canal. Harvard University Press, Cambridge, Mass.

- Also at Oak Ridge Reservation, the committee learned that a building at the K-25 facility (now the East Tennessee Technology Park) had been decontaminated up to eight feet from the floor, with the stipulation that no activities would be allowed above that height. It is not difficult to imagine, however, that the eight-foot limit eventually will be ignored or forgotten by users of the building. In addition, dust and dirt that slough from the walls above eight feet and the ceilings and other high structural features may contain contaminants.

These examples vary; the first is the most egregious; the last is speculative. Moreover, the well water examples illustrate both the limitations of institutional controls and ways that oversight can compensate for their deficiencies. Nevertheless, all of these examples suggest that institutional controls should not be relied upon for proper performance. Sidebar 5-2 gives another example of the fallibility of institutional controls.

> **SIDEBAR 5-2**
>
> **THE BIKINI ATOLL EXPERIENCE: INHERENT FALLIBILITY OF INSTITUTIONAL CONTROLS AND THE VIRTUES OF "DEFENSE IN DEPTH"**
> **(by A. Ballou Jennings and Thomas Leschine, University of Washington)**
>
> Between 1946 and 1962 the United States conducted 109 nuclear weapons tests in the Pacific Proving Grounds. Twenty-three of these tests were conducted on or near Bikini Atoll, located in the Marshall Islands, with a total yield exceeding 75,000 kilotons. Documentary photos and film footage of the 1954 *Bravo* shot, which vaporized three small islands and left a crater one-mile wide and 200 feet deep, have become iconic images of the era of large-scale nuclear testing that was ushered in by the Cold War. Bikini's 167 inhabitants agreed voluntarily to evacuate, with the understanding that the relocation was to be temporary. With little understanding at the time about the longevity of radiation effects in the environment, there appeared to be little reason to expect that the island could not be returned to habitability and resettled soon after the cessation of testing (Weisgall, 1994). In 1947 the United States became administrator of the Trust Territory of the Pacific, with obligations to promote "economic advancement and self sufficiency," to develop and regulate the use of natural resources, and to protect the health of the inhabitants.
>
> Atomic testing at Bikini continued until 1962. Following debris removal and the replanting of vegetation, Bikini Islanders began to resettle in 1972. Contamination risks were very much of concern. Bikinians were instructed to avoid eating locally grown foods and to limit their consumption of coconuts in particular. The importance of coconuts in the traditional Marshallese diet was not fully appreciated, however, and the required medical monitoring that was part of the resettlement agreement soon began to detect increasing body burdens for both strontium-90 and cesium-137. By 1978 the dose levels observed in many inhabitants far exceeded even the highest pre-settlement estimates (Robison et al., 1997). Reexamination of the scientific judgments that supported the resettlement revealed that the evidence available at the time on the behavior of radionuclides in soils was derived from continental soils and not the calcium carbonate-rich soils of Pacific islands, where cesium readily substitutes for potassium in plant uptake. Coconut trees

The weaknesses of different types of institutional controls have been discussed in various reports and articles (see, e.g., English and Inerfeld, 1999; Applegate and Dycus, 1998; Hersch et al., 1997; Pendergrass, 1996) and have been detailed in a draft reference manual on institutional controls by the U.S. Environmental Protection Agency (EPA) (1998a). Many of the weaknesses concern the fallibility of memory and the susceptibility of present-day intentions to future political and economic pressures. More broadly, as discussed in Chapter 7, there are concerns about institutional constancy, the atrophy of vigilance, and the problematic nature of follow-up and enforcement (LaPorte and Keller, 1996; Freudenburg, 1992), especially when stewardship impedes use for economic gain of desirable property that happens to be contaminated.

A study by the National Research Council (1995c) reported some serious concerns about the efficacy of long-term institutional controls for the proposed high-level waste repository at Yucca Mountain, Nevada. While Yucca Mountain is not a DOE "legacy site," findings about the reliability are relevant: "(1) institutional controls cannot be relied upon to protect a repository against intrusion, but (2) they should be used nonetheless as an added measure of protection" (see Sidebar 5-3).

Oversight and Enforcement

Oversight and enforcement activities, if carried out with continuous vigilance, should have the potential to help compensate for the deficiencies of other stewardship activities. The act of placing a use restriction on a parcel, proved to take up large quantities of cesium under the conditions that prevailed on Bikini, and became the primary pathway conveying radiological contamination to the human population.

In expectation that doses could exceed established radiological protection limits, follow-up radiological surveys and medical monitoring had been put into place. According to an internal memorandum in an archive compiled for public release by the U.S. Atomic Energy Commission (1978), "doses to resettled populations were expected to exceed dose limits." The idea was that the resettlement strategy could be adjusted as the true radiation exposure picture emerged. In this regard, the system put in place can be said to have worked. Errors that resulted from dose estimates, having been based on inappropriate models, were compensated for by medical and environmental monitoring able to detect an exposure problem in the human population and trace it to its sources in the environment.

Government officials had expected the resettled Bikinians to be aware of the risks of returning to their island home, to self-police their food consumption habits (despite little effort at risk communication), and to accept the risks of radiation exposure in exchange for the benefits of being able to return to their ancestral home. These assumptions proved to be in error. Bikini is judged safe for human habitation today, but with the proviso that all food and drink be supplied from outside. It thus remains unlikely that the traditional way of life the original Bikini evacuees expected soon to be able to resume can be reestablished for generations yet to come. Ironically, the conditions imposed by the need for continued radiological protection have proved well suited to a newer group of occupants, and Bikini has now become a major international tourist destination for underwater diving.

REFERENCES

Robison, W.T., K.T. Bogen, and C.L. Conrado. 1997. An updated dose assessment for resettlement options at Bikini Atoll—A U.S. nuclear test site. Health Physics 73:100-114.

U.S. Atomic Energy Commission. 1978 (April 3). Staff Review of Bikini Atoll Cleanup and Rehabilitation. U.S. DOE Archives, 326 U.S. Atomic Energy Commission, Radiological Survey, McGraw Collection, Washington, D.C.

Weisgall, J.M. 1994. Operations Crossroad: The Atomic Tests at Bikini Atoll. Naval Institute Press, Annapolis, Md. 440 pp.

---

zoning it, posting a sign, or erecting a fence will not be sufficient to assure that an ongoing site management measure will continue to perform as expected. An example of a problem in the short term was the rapid deterioration and disappearance of some of the signs warning of the contamination of East Fork Poplar Creek in Oak Ridge, Tennessee (Mulvernon, 1998); such a problem is all the more likely in the long term. Similarly, engineered systems to isolate contaminants will, as noted in Chapter 4, require maintenance and monitoring to ensure their long-term efficacy, and these ongoing management activities will, in turn, require oversight and, if necessary, enforcement.

Oversight and enforcement may be difficult to accomplish, however. For example, creating enforceable easements can be quite complicated. As indicated earlier, a property interest must be conveyed and recorded with the appropriate local government. Everyone with a property interest needs to be involved in agreeing to the easement. Parties such as a mortgage, lien, or earlier easement holder need to be involved along with the property "owner" during easement negotiations, because they have a legal interest in the property (U.S. Environmental Protection Agency, 1998a). Moreover, proprietary controls such as covenants and easements generally can be in force only if recorded and only by the party that holds the right to prevent restricted activity. If not enforced in a timely way, these controls will be extinguished by the courts; furthermore, such restrictions are not always binding on future property owners.

Local governmental ordinances also have serious weaknesses. Zoning is subject to change through exceptions, variances, and amendments, and even if zoned use restrictions remain in effect they may in some instances be overturned in court upon the appeal of a property owner. Other ordinances, such as building codes and well-

SIDEBAR 5-3

INSTITUTIONAL CONTROLS AT YUCCA MOUNTAIN GEOLOGICAL REPOSITORY
(by Thomas A. Cotton)

In Section 801 of the National Energy Policy Act of 1992 (P.L. 102-486), the U.S. Congress directed the U.S. Environmental Protection Agency (EPA) to promulgate standards to ensure protection of public health from high-level radioactive wastes in a deep geologic repository that might be built under Yucca Mountain in Nevada. Congress explicitly asked the National Research Council to advise EPA on the technical bases for such standards, including addressing the question of the efficacy of long-term institutional controls of a high-level radioactive repository at Yucca Mountain. In its report, the National Research Council Committee on the Technical Bases for Yucca Mountain Standards (TYMS) found that, while a Yucca Mountain repository would not be a "legacy site" (National Research Council, 1995), the issues surrounding the role of institutional controls are similar and the TYMS committee's findings about reliability are relevant to the Department of Energy legacy waste sites discussed in the current report. In its report, the TYMS Committee took a dim view of the efficacy of institutional controls:

> With respect to the second question of Section 801, we conclude that it is not reasonable to assume that a system for post-closure oversight of the repository can be developed, based upon active institutional controls, that will prevent an unreasonable risk of breaching the repository's engineered or geologic barriers or increasing the exposure of individual members of the public to radiation beyond allowable limits. This conclusion is founded on the absence of any scientific basis for making projections over the long term of the social, institutional, or technological status of future societies. Additionally, there is no technical basis for making forecasts about the long-term reliability of passive institutional controls, such as markers, monuments, and records. (National Research Council, 1995, p. 11 and 105).

In their explanation for this conclusion, the committee stated:

> For some initial period, human intrusion could be managed through active or passive controls. As long as they are in place, active institutional controls such as guards could prevent intruders from coming near the repository. We conclude, however, that there is no scientific basis for making projections over the long term of either the social, institutional, or technological status of future societies. Relying on active controls implies requiring future generations to dedicate resources to the effort. There is, however, no scientific basis from which to project the durability of governmental institutions over the period of interest, which exceeds that of all recorded human history. On this time scale, human institutions have come and gone. We might expect some degree of continuity of institutions, and hence of the potential for active institutional controls, into the future, but there is no basis in experience for such an assumption beyond a time scale of centuries. Similarly, there is no scientific basis for assuming the long-term effectiveness of active institutional controls to protect against human intrusion. Although it may be reasonable to assume that a system of post-closure oversight can be developed and relied on for some initial period of time, there is no defensible basis for assuming that such a system can be relied on for times far into the future. Between these limits, the ability to rely on such active institutional systems presumably diminishes in a way that is intrinsically unknowable. We have seen no evidence to support a claim to the contrary. People might disagree, of course, on their predictions for how long into the future active institutional controls might survive and remain effective. (National Research Council, 1995, p. 106)

However, the committee went on to observe that:

> although there is no scientific basis for judging whether active institutional controls can prevent an unreasonable risk from human intrusion, we think that if the repository is built such controls and other activities can be helpful in reducing the risk of intrusion, at least for some initial period of time after a repository is closed. Therefore, although it cannot be proven, we believe that if a repository is built at Yucca Mountain, a collection of prescriptive requirements, including active institutional controls, record-keeping, and passive barriers and markers, will help to reduce the risk of human intrusion, at least in the near term. The degree of benefit is likely to decrease over time. Further, once other knowledge of the repository is lost, passive markers could attract

> the curious and actually increase the risk of intrusion. Nonetheless, we conclude that the benefits of passive markers outweigh their disadvantages, at least in the near term. (National Research Council, 1995, p. 108)
>
> In summary, the TYMS committee concluded that (1) institutional controls cannot be relied upon to protect a repository against intrusion, but (2) they should be used nonetheless as an added measure of protection. (Section 801 of The National Energy Policy Act of 1992 in fact requires perpetual post-closure oversight of a repository.) This is compatible with the philosophy adopted in EPA's original regulations for high-level waste and transuranic waste repositories (40 CFR part 191), which precluded reliance on institutional controls for more than 100 years but at the same time required continued control for as long as feasible. Such an approach does not allow institutional controls to be used as a way to avoid doing the best job to isolate the waste that is technically possible and financially feasible, but instead views institutional controls as providing redundant protection. This appears to be consistent with the conclusion of the present report that stewardship measures are the least favored of the three legs of the stool, but nonetheless have a role to play in any long-term plan that must leave some contaminants at the site in question.
>
> **REFERENCE**
>
> National Research Council. 1995. Technical Bases for Yucca Mountain Standards. Committee on the Technical Bases for Yucca Mountain Standards, National Academy Press, Washington, D.C.

drilling restrictions, may not be enforced effectively by local government agencies, especially if the agency is under-staffed, lacks specialized technical competence, or becomes preoccupied with other responsibilities over time.

In the words of the draft EPA guidance document on institutional controls (U.S. Environmental Protection Agency, 1998a, p. 52):

> . . . both proprietary and governmental controls have weaknesses in terms of long-term reliability. . . . Where turnover in ownership is likely, common law doctrines restricting enforcement by parties who do not own adjoining land can render proprietary controls ineffective; governmental controls may be preferable in such cases. At the same time, over the long term governmental controls may not be effectively enforced because political and fiscal constraints may influence a State or local government's exercise of its police power.

A draft DOE study (U.S. Department of Energy, 1997a, p. 2-8) states the problem bluntly: " . . . there is little or no evidence demonstrating the effectiveness of enforcing and maintaining institutional controls." A study by the National Research Council (1999a) reaches a similar conclusion, noting that land-use controls "for both legal and physical reasons, are very difficult to enforce."

For a set of institutional management activities to remain effective, there should be both the means to detect impending or actual failure and also the authority and will to require those responsible to correct the problem. Unfortunately, as illustrated by the problems with use restrictions, oversight and enforcement activities can have important limitations, particularly if they are not conducted with a clear allocation of responsibility and authority as well as adequate training and funding. Moreover, even if an oversight arrangement has been fairly effective in the short run (the DOE Oversight Office within the Tennessee Department of Environment and Conservation, which operates with funding provided by DOE, might be one example—see later in this chapter), there is no guarantee that the present arrangement will provide the expected vigilance for decades or centuries into the future.

Information Collection, Storage, and Retrieval

To be useful, information must be carefully and systematically collected and stored, and information from the receding past must remain easily retrievable. Such an information management program requires stable financing and an administrative mechanism that is maintained over the entire period that the information is needed. Information management is proving to be more challenging and potentially troubling than policymakers sometimes expect (see ICF Kaiser, 1998).

One important, informal way to respond to some information challenges is to take advantage of the knowledge of present and past workers who have carried out the day-to-day jobs of weapons production and waste disposal. These employees have sometimes been treated as adversaries, or "whistleblowers," yet they can be a fount of knowledge about the existence and location of wastes. One example is the recent revelation concerning the dearth of trustworthy information available to workers at the Paducah (Kentucky) Gaseous Diffusion Plant that would have informed them about potentially harmful radioactive contamination in their work environment through the many years of the plant's operation (*The Washington Post*, September 21, 1999, and December 23, 1999). Nevertheless, surprises continue to occur. For example, members of the Site Specific Advisory Board at Rocky Flats, Colorado, pointed out to the committee that during the excavation in late 1998 of a second waste disposal trench at Rocky Flats, the collapse of a trench wall revealed a buried waste container that had not previously been known. As the recollection or availability of current and past weapons facility workers declines, the problem of fading institutional memory is likely to worsen.

The more formal process of information storage and retrieval needs attention as well. Today's information management systems, which generally are computerized, in principle, can offer vast improvements over paper-based record-keeping systems, such as ready access to information that might have been virtually impossible to locate in reams of paper records, data integration through means such as geographic information systems, and, at least in principle, the potential for accessibility by citizens as well as employees.

Unfortunately, over the long haul, computer-based information management systems may be much less legible and durable than paper-based systems. Information professionals such as archivists—often heavy computer-users themselves—have expressed concerns about the long-term reliability of computer-based information systems (Tangley, 1998). Computer-based records can lose their accessibility relatively quickly if advances in computer technology make prior means of storing information (e.g., computer tapes or 5-1/4-inch diskettes) obsolete. Even more fundamentally, there is the problem of magnetic degradation over time, a problem that, by some estimates, can take place within relatively few years. The very long time frames over which institutional management must remain reliable at DOE sites presents special problems for the information storage and retrieval systems upon which DOE site stewards must depend. The durability beyond even a decade or so of some information storage media currently in heavy use (VHS recording tape and CD-ROMs) has been questioned (Tangley, 1998). Even when recording tapes and disks remain intact, the hardware and software needed to read them may no longer be available.

Site information needs to be maintained in accessible paper or electronic forms for as long as the site must be protected. It will need to be systematically, and probably repeatedly, transferred to new electronic data bases or other systems before its present form deteriorates or becomes obsolete. As with retrieving information from past nuclear weapons plant workers, the ability to readily retrieve information from past computer-based systems may become increasingly difficult.

Information Dissemination

Still another problem arises with ongoing information dissemination. For successful institutional management of risks, relevant information should get to the people who need it, when they need it. These people may be those responsible for managing or overseeing the site, or they may be citizens potentially affected by the site. While an aggressive information outreach campaign may be mounted when remedial actions are being decided and conducted, will it remain aggressive for decades to come? Moreover, as the nature of the site's risks change—for example, if off-site migration of contaminants occurs—the contents and targets of an information dissemination campaign will need to be adjusted accordingly.

Periodic Reevaluations of the Site Protective System

DOE is responsible for the CERCLA five-year reviews at its sites (Executive Order 12580, *Superfund Implementation*). Thus, every five years at a minimum, an assessment should be done by DOE of site institutional management systems. The concern is that even if these assessments are legally required, they may not be carried out. For example, the U.S. Environmental Protection Agency (EPA) is responsible for completing the CERCLA five-year review at private sites. However, as of March 31, 1999, 143 five-year reviews were overdue, with an average overdue period of 17 months, and the backlog of uncompleted reviews had increased threefold since the previous audit in 1995 (U.S. Environmental Protection Agency, 1999), an indication that even a legally binding obligation can fail.

In its guidance on the use of institutional controls when federal agencies transfer property to non-federal users, EPA headquarters states "Even if implementation of the institutional controls is delegated in the transfer documents, the ultimate responsibility for monitoring, maintaining, and enforcing the institutional controls remains with the lead federal agency responsible for cleanup" (U.S. Environment Protection Agency, 2000). In addition, some EPA regional offices have issued their own policy statements regarding measures to assure the long-term effectiveness of institutional controls or land use controls at residually contaminated federal property. In these policy statements, the regions call for the federal facilities within their jurisdictions to specify clear plans for implementing, enforcing, and monitoring institutional or land use controls whenever a site to be remediated will have residual contamination necessitating use restrictions (U.S. Environmental Protection Agency, 1998c).

Despite these guidance documents, it would appear that systematic and comprehensive re-evaluations by DOE of a site's protective system, including but not limited to its institutional controls, are by no means guaranteed, especially over time. As will be discussed in Chapter 7, individuals and organizations often cut back on tasks regarded as "peripheral" when pressed for time or money. Periodic reevaluations could easily be deemed peripheral over the long term. They are even more likely to be regarded as peripheral if the site has been leased to or acquired by another party.

Developing Remediation Options and Understanding of Contaminant Behavior

Similarly, promoting development of new options to decontaminate further a remediated but residually contaminated site—or directed research in ways that resolve scientific uncertainties that could compromise the effectiveness of site protective measures—could quite easily "fall through the cracks," especially if other priorities are clamoring for attention. Pursuing research and development (R&D) to improve characterization of waste and the contaminated environment and to provide new decontamination options and more effective means to stabilize, isolate, and monitor contamination, may also be given short shrift, especially if the current downward trend in funding for DOE remediation technology R&D continues (National Research Council, 1999b).

Deficiencies in site and waste characterization at DOE sites, in the scientific and technical understanding and tools available to support this work, and in the technologies available to address site contamination problems, have been pointed to repeatedly (National Research Council, 1995b; 1997a,c; 2000b). The ramifications of the resulting gaps in scientific understanding of site and waste conditions potentially loom larger as intensive site remediation and strong DOE and contractor presence at sites give way to less intensive on-site work and reduced levels of on-site presence. Such gaps in understanding include knowledge of the behavior of residual contaminants in site environments, the removal efficiency of site remediation that has already been completed or is ongoing, and the performance of barrier technologies in use to contain residual site contamination. In commenting on recent drastic revisions in estimates of the travel time of subsurface contaminants at the Radioactive Waste Management Complex at the Idaho National Engineering and Environmental Laboratory (INEEL), long held to be on the order of tens of thousands of years, but very recently revised to only a few tens of years, that report notes (National Research Council, 2000b, p. 30, Sidebar 2.6):

The uncertainty of these estimates is attributed to several factors, including incorrect conceptualizations of the hydrogeologic system, improper simplifying assumptions, incorrect transport parameters, and overlooked transport phenomena.

CHARACTERISTICS OF AN EFFECTIVE STEWARDSHIP PROGRAM

The overarching requirement for an effective stewardship program is that it be *reliable*. A reliable program has a reasonable likelihood of achieving its objectives over the period it must remain in effect. For example, the reliability of a stewardship program will be enhanced if the characteristics of *layering* and *redundancy* are applied. These and other related characteristics are discussed below.

- ***Layering and redundancy***. Layering and redundancy rest on the notion that a stewardship program is more likely to be reliable if it avoids "putting all its eggs in one basket." For purposes of this report, *layering* means using several measures to carry out roughly the same function (e.g., several institutional controls to restrict uses); *redundancy* means creating a situation in which several entities are responsible for or have a vested interest in the effectiveness of the measures. Both the layering and redundancy should be carefully designed to assure that the relevant organization will have appropriate incentives to protect public health and safety. If properly implemented, however, the layering and redundancy concept can be comparable to that of using several different physical barriers to increase the robustness of a contaminant isolation system.
- ***Ease of implementation***. A stewardship activity must be capable of being put into effect, and it also should be reasonably easy to keep in effect.
- ***Monitoring commensurate with risks***. Monitoring methods and schedules need to be commensurate with the harm that could be caused in the case of release of contaminants or failure of a monitoring system. The monitoring strategy should also include indicators that trigger modification or termination of the activity based on changes in risk to human health or the environment.
- ***Oversight and enforcement commensurate with risks***. As discussed above, one key stewardship activity is to have a "watchdog" over other stewards and stewardship activities. For the watchdog to be effective, however, it must have teeth. For example, if DOE leases property to a private party on the condition that no construction can occur without prior DOE approval and the tenant then violates this condition, the federal government must be willing and able to sue the tenant for damages and termination of the construction project, and possibly the lease.
- ***Appropriate incentive structures***. Given that different people and institutions respond to different incentives, attention needs to be devoted to assuring that site stewardship managers will be appropriately motivated for carrying out the needed tasks over time, not only in implementing and monitoring an institutional management plan, but also in the vigilant safeguarding of remaining hazardous and radioactive materials. With careful planning it may be possible to identify or develop institutional managers having clear incentives to act in ways that preserve stewardship systems. Certain types of local citizen groups that have clear concerns over public health might be expected to have such incentives. Recent research has suggested that the Regional Citizens Advisory Councils that were set up (and provided with reasonably stable funding) after the *Exxon Valdez* oil spill in Alaska do indeed seem to be playing an important sentinel function at least for a period of a decade, working against the kind of "atrophy of vigilance" that had been seen prior to the spill (Busenberg, 1999; Freudenburg, 1992; Galanter, 1974).
- ***Adequate funding***. Implementing, monitoring, and appropriately modifying stewardship activities will require adequate and reliable financial resources throughout the activities' required lifetimes. It is not clear whether regional citizens advisory councils, for example, will be able to exercise the same degree of influence in the absence of reasonably stable funding.
- ***Durability or replaceability***. A stewardship activity should endure either for as long as the site's residual contaminants remain hazardous, or until the activity can be refreshed or replaced by an equally reliable substitute activity. For example, as discussed above, institutional controls such as zoning restrictions may not survive long; if they do not, they need to be succeeded by other use restrictions appropriate to the remaining risks. As with contaminant isolation technologies, stewardship activities at many DOE waste sites will need to be effective for

much longer time periods than our experience to date with them. The associated uncertainties underscore the need to develop ways to improve their reliability over time.

FUTURE DIRECTIONS FOR IMPROVING STEWARDSHIP

If a site's residual contaminants present risks to human or environmental health and safety, stewardship—including but not limited to institutional controls—will be required. Understanding the current limitations of various stewardship activities can lead to developing possible approaches for improving stewardship. A few approaches are briefly discussed below. These approaches are presented here as possibilities only; they are not necessarily endorsed as preferred solutions. In addition, these approaches are not necessarily mutually exclusive; instead, some are broad while others address a particular current problem.

Stewardship Entity

One possible approach is to identify or create a single entity with primary responsibility for maintaining and enforcing stewardship activities. Its mandate should be clearly defined. To fulfill this mandate, it would need legal authority and responsibility (including appropriate susceptibility to sanctions if the entity were to be derelict in its duties), as well as stable funding (E. Frost, Attorney, Leonard, Hurt, Frost & Lilly, presentation to a group from the committee, June 9, 1999; Probst and McGovern, 1998).

Activities of the entity might include all of those listed previously in this chapter under "Components of a Comprehensive Stewardship Program." The entity might take title to sites (public or private), lease or transfer property for reuse and retain the proceeds, support research to advance stewardship activities as well as contaminant reduction and isolation, and train and use local citizens, organizations, and businesses to perform monitoring and maintenance. Such an entity would take advantage of local knowledge of the site and its surroundings, resulting in improvement of the local economy and increasing awareness of site use restrictions. The entity might be subject to citizen suits for failure to carry out its responsibilities. It might be funded by a trust, by Congress, and/or by site lease and sale proceeds. Such an entity might be an organization such as an existing federal or state agency; alternatively, it might resemble a trust.

Funding by Congress may be a highly questionable proposition, in view of the failure (so far) of the Nuclear Waste Fund mechanism, which was created to provide assured funding for the development of a permanent high-level waste repository. The last 18 years' experience with the Nuclear Waste Fund shows that the federal budget system is ill suited to that sort of effort. The Nuclear Waste Policy Act of 1982 required DOE to enter into contracts with nuclear utilities, committing DOE to begin acceptance of their spent fuel by January 31, 1998, in exchange for payment of an annual fee of 1 mill per kilowatt hour. This fee brings in over $600 million annually to the federal treasury, yet Congress has been appropriating less than 1/3 of that amount each year for development of a repository, and has failed to provide the funds requested by DOE for the program for a number of years. Since all of the budget control laws apply to the appropriations from the Nuclear Waste Fund into which the fee is paid, expenditures from that Fund for the repository program are constrained despite the high annual income to the Fund and the legal obligation to provide disposal services.

Trusts

A trust is a legal entity that holds an asset (money or property) for the benefit of beneficiaries (see Sidebar 5-4). Trustees are designated to manage the asset and are legally obligated to manage it in the best interests of the beneficiaries. Beneficiaries can sue for damages or injunctive relief if the trustees violate their fiduciary responsibilities. In a trust for a DOE waste site, the federal government would create the asset (money and/or property), the state or region might be the beneficiary, and the trustees might be either individuals appointed for multi-year terms or an entity that holds title to the site subject to use restrictions. The beneficiary could sue the trustees in federal court for violations of their fiduciary responsibilities. Several examples of trusts for contaminated sites follow:

1. Trusts are used in the RCRA program to ensure funding for post-closure care of non-federal hazardous waste management sites: Site owners and operators create the asset (money), a bank is the trustee, and the state is the beneficiary.

2. Under the Presidio Trust Act of 1996 (16 U.S.C. §460bb appendix; enacted as Title I of H.R. 4236, P.L. 104-333, November 12, 1996; and amended by P.L. 105-83, November 14, 1997), the parts of the Presidio (a former Army post near the Golden Gate Bridge in San Francisco, California) not retained by the U.S. Department of the Interior were put under the responsibility of a wholly owned government corporation. It manages the leasing, maintenance, and improvement of the Trust properties, it can negotiate and enter into agreements, leases, and contracts to carry out its functions, and it develops rules and regulations governing its operation. It can retain proceeds received by the Trust for the administration, maintenance, improvement, etc., of the properties, but it may not sell or otherwise convey the title to these properties. It also can sue and be sued to the same extent as the federal government.

3. The State of Tennessee Department of Environment and Conservation signed a consent order on October 29, 1999, that requires DOE to make yearly payments of $1 million for 14 years into a trust fund maintained by the State of Tennessee to cover post-cleanup monitoring costs at the Oak Ridge Reservation disposal facility (see Sidebar 5-4). The consent decree may become a model for other states when entering into agreements to allow DOE to dispose of waste on site to obtain funds for long-term monitoring. The expectation is that, after 14 years, the trust fund should generate enough interest to cover the expected yearly operation and maintenance costs for the facility (about $650,000, according to *Inside Washington Superfund Report*, November 10, 1999, p. 8). A limitation of the Tennessee Perpetual Care Trust Fund is that the federal government can make no financial commitments beyond one year under the Anti-Deficiency Act. Congress could and should address longer-term funding issues for such trusts.

Transferring Partial Authority and Responsibility to Other Federal or State Agencies

Another alternative for authoritative and responsible management of sites under stewardship is to combine state or federal agencies with the organizations responsible for the contamination. The Rocky Mountain Arsenal in Colorado provides an example of this approach. The U.S. Department of Defense (DOD) and Shell Oil Company are the liable parties for remediation of the Rocky Mountain Arsenal. Congress directed DOD to transfer jurisdiction of certain portions of the Rocky Mountain Arsenal to the U.S. Department of the Interior for management as part of the National Wildlife Refuge System. Management of the transferred property remains subject to any necessary cleanup activity; DOD is responsible for the cleanup and is liable, under CERCLA, for future cleanup activity. The Rocky Mountain Arsenal National Wildlife Refuge Act of 1992 (Public Law 102-402) requires that the real property that is exempted from the transfer but subsequently disposed shall be subject to deed restrictions prohibiting in perpetuity residential or industrial use, groundwater use, hunting and fishing for consumptive use, and agriculture. Given that this measure is still a relatively recent one, its long-term reliability is unknown.

Another approach would be one of four kinds of public or quasi-public institutions that have had relatively successful track records in safeguarding materials over long periods of time; for example, libraries, archives, museums, and, at least for the last 100 years, the U.S. National Park Service. All four share at least a pair of characteristics that may be noteworthy in the present context. First, in most cases, they are not expected to balance preservation with economic development, instead being given clear responsibility for preservation duties. Second, all four are expected to carry out their preservation duties in ways that permit or even encourage controlled public access. The potential for public access appears to increase the visibility of an organization's performance to members of the broader public, who then can be expected to have an interest in the constancy of the institution's vigilance over time (Busenberg, 1999; Clarke, 1993; Freudenburg, 1992; LaPorte and Keller, 1996; Shrader-Frechette, 1993).

Remediation Easement

As noted previously, conservation easements have become reasonably widespread and are recognized by

SIDEBAR 5-4

TRUST FUNDS AND INSTITUTIONAL MANAGEMENT
(by Elizabeth K. Hocking)

One of the keys to the success of an institutional management plan is funding that is adequate and consistent throughout the required life expectancy of the plan. Federal facilities are currently funded for one year of operation at a time. Funding for one fiscal year could be dramatically increased or decreased for the next year depending upon congressional findings and appropriations. A one-year funding cycle is incompatible with achieving the goals of a multi-year institutional management plan. An irrevocable trust for institutional management plans should be evaluated as a possible solution to this funding dilemma.

A trust is a legal entity that holds an asset (money or property) for the benefit of beneficiaries. Trustees are designated to manage the asset and obligated by law to manage it in the best interests of the beneficiaries. Beneficiaries can sue for damages or injunctive relief if the trustees violate their fiduciary responsibilities. Trusts are presently used in the Resource Conservation and Recovery Act of 1976, as amended (RCRA) program to ensure adequacy of funding for post-closure care of non-federal hazardous waste disposal sites. Site owners and operators create the asset (money), the state is the beneficiary, and a bank is the trustee. In an institutional management trust, the federal government would create the asset (money and/or land), the state could be the beneficiary, and the trustees could be individuals appointed for multiple year terms or an entity that actually holds title to the land subject to the institutional management plan. The beneficiary could sue the trustees in federal court for violations of their fiduciary responsibilities.

Creation of an institutional management trust fund raises several questions. First of all, how can the federal government commit itself to an irrevocable trust? How would the legal document establishing the trust be constructed to preclude future congresses from disestablishing the trust or under-funding it? Second, what is the intended use of the trust? Should the trust be used for operations and maintenance related to the institutional management plan? Should it be used only if the U.S. Department of Energy (DOE) or a successor agency fails to implement the institutional management plan and damages arise (human health or environmental degradation)? Third, how would the amount of the asset that needs to be held in trust be determined and how could the asset be replenished? Fourth, would it be possible to establish trust arrangements only for specific sites, rather than for the DOE complex as a whole, to take into consideration such differences between sites in level and type of contamination, degree of remediation that has been accomplished, and anticipated residential and industrial land use.

In most cases, the life-cycle cost of the operations and maintenance of the institutional management plan will be difficult to determine with certainty at the time the trust is created. Furthermore, the immediate deposit in the trust of the reasonably expected life-cycle cost of the plan for the entire DOE complex could cause a dramatic and unhealthy increase in the federal budget. How should these issues be addressed? If the trust is to be used only upon failure of the DOE or a successor agency to comply with the institutional management plan, how would the initial dollar amount of the asset be determined? How would the trust be replenished if original cost estimates were inaccurate? Fourth, what would be the obligations and rights of the trustees? Can the trustees allow re-use of the land, and who would establish the conditions of re-use? Will the trustees be exempt from suits brought by beneficiaries if Congress has under-funded the trust?

AN EXAMPLE: THE TENNESSEE PERPETUAL CARE TRUST FUND

Pursuant to the Tennessee Hazardous Waste Management Act (Tenn. Code Ann. § 68-212-101 et seq.), the Commissioner of the Tennessee Department of Environment and Conservation has the authority to require the payment of sums to a statutorily created fund called the "Perpetual Care Trust Fund" if the Commissioner determines that there is a reasonable probability that a site "will eventually cease to operate while containing, storing, or otherwise treating hazardous waste on the premises that will require continuing

(continued)

> **SIDEBAR 5-4 (Continued)**
>
> and perpetual care or surveillance over the site to protect the public health, safety, or welfare." On October 29, 1999, the Commissioner made such a determination for the U.S. Department of Energy (DOE) Oak Ridge Reservation and signed a consent order decreeing that DOE shall pay the Tennessee Department of Environment and Conservation (TDEC) the sum of $14,000,000, payable in 14 annual installments.
>
> These funds are to be deposited into a "Perpetual Care Trust Fund" for use by TDEC for its performance of the surveillance and maintenance of the Environmental Management Waste Management Facility at the Oak Ridge Reservation, Oak Ridge, Tennessee. The surveillance and maintenance is to begin upon completion by DOE of the disposal of contaminated media and radioactive and hazardous wastes in an engineered, above-grade, earthen disposal cell, and construction of a RCRA-compliant cap to cover the cell and of associated monitoring systems. Unlike private trusts, the TDEC trust will remain under the control of a government entity, albeit in this case a state instead of the federal government. As such, the arrangement is likely to provide a test of such trusts, as well as providing tentative answers to the questions posed above.

many states, offering options that are worth considering. Alternatively, it may be possible to create a federal remediation easement that overcomes the possible enforcement problems of conventional easements and allows for broader usage than conservation easements. Such an easement could be patterned after the hazardous substance easement described in CERCLA reauthorization and amendment bills introduced in the 105th U.S. Congress (H.R. 3000, Superfund Reform Act; H.R. 2750, Superfund Cleanup Acceleration and Liability Equity Act; and H.R. 2727, Superfund Acceleration, Fairness, and Efficiency Act). The hazardous substance easements proposed in these bills were enforceable for 20-year periods with additional 20-year renewal periods. They were enforceable against all owners and subsequent purchasers as well as all holders of interest in the property regardless of whether the interest was recorded or not. The easement, as described in H.R. 3000, could be assigned to "a State or other governmental entity that has the capability of effectively enforcing the easement over the period of time necessary to achieve the purposes of the easement."

Insurance

Another approach might be to require that recipients of previously contaminated federal property have insurance against contamination liability. As it now stands, if newly discovered contamination is shown to have been caused by the federal government in the past and the property recipient did not contribute to the contamination, the recipient can seek recourse against the federal government. Perhaps this should be insurance enough, but it does create a potentially undesirable incentive system in that future government officials might well have few resources for dealing with the contamination, but large resources for resisting action.

Private-sector insurance mechanisms deserve greater attention in connection with privatization options. One of the concerns that citizens have expressed about having private-sector firms take over the responsibility for decontamination is that such firms might have incentives to "cut corners," potentially endangering public, worker, and environmental health and safety. One possible approach for minimizing such undesirable incentives might be to include strict liability provisions and strong legal safeguards for local residents, along with the requirements that the private firms obtain and maintain liability insurance for their management activities. So long as all relevant parties in advance know these provisions, this could produce a more desirable incentive structure. Moreover, if the property recipient were required to have insurance against contamination liability, the insurance company would have a vested interest in initially conducting an independent appraisal of the risks associated with property transfer and ensuring that any use restrictions are observed. Operating companies' insurance premiums would be lowered in cases where decontamination efforts were more effective, insurance companies would share the incentives to control costs and to improve performance, and government bodies could be placed in the position of being relatively impartial arbiters of the interests of other parties rather than of having the potential conflict of interest of needing to minimize governmental costs and liabilities as well as the remediation of contamination.

RELEVANT RESEARCH AND DEVELOPMENT NEEDS

The previous two chapters addressed the need to improve contaminant reduction and isolation technologies. Such improvements, however, will not in themselves lead to reliable long-term site institutional management unless gaps in basic scientific understanding are also addressed. These gaps include, for example, deficiencies in our ability to make accurate estimates of subsurface contaminant behavior, especially in the conceptual understanding of this behavior to enable accurate and robust modeling.

Our understanding of how to develop and implement stewardship also must be improved, especially with respect to the appropriateness and reliability of stewardship activities. Improvements through new research are needed in the following areas:

- investigating ways to make existing stewardship activities more effective;
- developing new institutional controls (e.g., the federal remediation easement mentioned above);
- designing new, more effective and efficient systems for monitoring and oversight;
- evaluating the characteristics of organizations best suited to take responsibility for stewardship activities; and
- developing methods to predict and compare the effectiveness of alternate stewardship approaches.

If stewardship responsibilities are to be vested in a single entity, research might be conducted on the following questions:

- What organization structure would be optimal? For example, would the entity be a private-sector firm with government oversight, a wholly owned government corporation, a government agency, or a quasi-governmental agency? Is an agency such as DOE, with its history of weapons production, more or less suited for a stewardship function than another agency with a different history and culture?
- What would the entity's property-related powers and responsibilities be? For example, could it hold private and public land? Lease property? Convey fee titles? Would it be bound by the existing property disposition protocols applicable to federal land?
- What would the entity's fiscal powers and responsibilities be? For example, could it commingle congressional appropriations with proceeds from leases or property transfers? Could it charge a maintenance and operation fee for federal government lands as well as the privately held lands turned over to it? How would that fee be determined?
- What incentives (or sanctions) would be needed to encourage governmental and private organizations to turn over stewardship of residually contaminated sites to the entity, and to motivate the entity to carry out its responsibilities?
- What roles would individuals and other organizations (e.g., members of affected communities, regulators) have?

As these questions make clear, even a potentially attractive "answer" such as a single stewardship entity leaves many issues still to be resolved. Moreover, it is not necessarily the right answer in all cases. Instead, as the following chapter helps to illustrate, contextual factors need to be taken into account in making decisions about the long-term disposition of the DOE waste sites.

6

Contextual Factors

As noted in Chapter 2, numerous contextual factors can affect the nature and extent of the measures taken to accomplish long-term institutional management. In particular, seven factors often constrain the range of decisions and actions realistically available:

- risk;
- scientific and technical capability;
- institutional capability;
- cost;
- laws and regulations;
- values of interested and affected parties; and
- other sites.

The measures of institutional management—contaminant reduction, contaminant isolation, and stewardship—were described in Chapters 3, 4, and 5, respectively. At any stage in the long-term disposition of a waste site, the above factors will affect how each of the three sets of measures (or "legs of the stool") is implemented, and also what the balance among the measures will be. These seven contextual factors thus can be thought of as the rungs of the committee's conceptual stool. For individual sites, given their variability, different emphasis may be placed on each of these contextual factors, depending on the contaminants present, current and projected future land use for the site and adjacent areas, and local and national economic, social, legal, and political considerations. These seven contextual factors and their characteristics and potential effects on site disposition decisions are considered below.

RISK

The primary objective in the disposition of most sites is to reduce the level of risk[1] to acceptable levels. Often, *human health risks* are of greatest concern. These risks can be categorized using dimensions such as the age of

---

[1] Risk is defined as the probability that something (a hazard) will cause harm or injury, combined with the potential severity of that harm or injury.

those at risk (e.g., adults, children), their relationship to the site (e.g., site workers, members of the public), the possible diffusion of the risk (e.g., local, global), and the nature of the possible effects (e.g., mortality, morbidity). Increasing consideration also is being given to *ecological risk* (i.e., the possibility of adverse impacts from contaminants on living organisms other than humans). While radiological protection standards for human health are thought to be protective of other living organisms in most cases (United Nations Scientific Committee on the Effects of Atomic Radiation, 1996; National Council on Radiation Protection and Measurement, 1991; International Atomic Energy Agency, 1976, 1979), some non-human species are particularly sensitive to certain chemical contaminants (e.g., copper and zinc concentrations acceptable in drinking water for humans are toxic to trout). In addition, disruptions to these organisms and their habitats from remediation activity or from prospective site reuse is also of concern.

Of the seven contextual factors listed above, risk is arguably the most important in site disposition as it, or perceived risk, may drive both the need for remediation and the level of stewardship required. The greater the risk, the greater should be the efforts required to reduce contaminants, isolate them, and carry out stewardship activities on sites containing residual contaminants. Further, the extent to which risk can be reduced often defines the extent of reliance on the respective "legs of the stool." Factored into this equation, however, some contaminant reduction and isolation measures also create human risks (e.g., by exposing remediation workers or by disturbing contaminants and making them mobile), or, as noted above, ecological risks. For example, contaminated sediments may be left in place in White Oak Creek at the Oak Ridge Reservation, in part to avoid disruption of the creek's ecology by dredging. Similarly, at the Nevada Test Site managers noted concerns that the surface soil cleanup could disrupt the site's sensitive desert ecology.

Risk and Performance Assessment

Risk is often estimated through *risk assessment*, essentially an attempt to estimate the hazards of contaminants to the environment and to various human populations, including sensitive groups such as children, the elderly, and pregnant women, and uncertainties associated with these estimates. From this process, the likely probability and consequences of adverse effects from a contaminated site, both as it presently exists and at some future, desired state, are assessed. (For detailed discussions of risk and risk assessment, see reports issued by the National Research Council [1983; 1989; 1994a,b; 1996a].)

A risk assessment, therefore, is (or should be) a comprehensive assessment of the entire system of measures to reduce, isolate, or otherwise limit exposure to site contaminants. In contrast, a *performance assessment* is more limited in scope, usually referring to an evaluation of whether a system satisfies predetermined design or performance criteria. As such, it contributes to assessing technical capability, discussed below.

Risk assessments typically use mathematical models that seek to represent how various factors interact to determine risk. Performance assessments similarly aim to estimate the performance of controls intended to limit risk exposure. Information is fundamental to either type of model in that it permits realistic estimation of model parameters and helps to determine a model's conceptual and mathematical structure and the appropriateness of its simplifying assumptions. For example, the computer model RESRAD (see Appendix G) is often used in both risk and performance assessments to estimate the direct exposure to radiation at DOE sites. The model incorporates assumptions of environment homogeneity that may or may not be appropriate to the particular waste and site conditions to which it is being applied. Chapter 7 and Appendix G provide more details on the capabilities and limitations of mathematical models in addressing site risks.

Uncertainty

As noted above, an important aspect of risk and performance assessment is *uncertainty*. Despite the desirability of having a high degree of confidence, uncertainties often arise, involving factors such as the following:

- ***Present condition of contaminants.*** The present identity, amount, form, and distribution of contaminants often is uncertain, especially when access to contaminants (e.g., subsurface contaminants) is limited to sampling and non-invasive techniques. Recent examples are the unexpected migration of plutonium (possibly in colloidal form) in groundwater at the Nevada Test Site (Kersting et al., 1999) and appearance of cesium-137 at the bottom of a 125-foot well in the Hanford Site (Rust Geotech, 1996). At the Nevada Test Site, there is considerable uncertainty as to the consequences of underground nuclear testing, including uncertainty about (a) the amount of contamination that now resides in groundwater, (b) the amounts and types of contaminant residues in the source term, (c) the amount and rate at which the contamination is mobilized by groundwater, and (d) the pathway(s) that the contamination may follow in the groundwater and the rates and concentrations associated with possible contaminant migration (see Appendix F). However, addressing these areas of uncertainty can raise new concerns. At the Hanford Reservation, for example, there has historically been great reluctance to drill additional bore holes that could help establish more accurately the extent of tank farm leakage for fear that such drilling could create new flow paths for subsurface contamination, exacerbating the condition of greatest concern (Conaway et al., 1997).

- *Future behavior of contaminants.* Contaminants can migrate (typically through soil, air, or water, but also through the reuse of contaminated materials), and they sometimes move through complex ecological cycles that may involve numerous species of flora and fauna. As elaborated in Chapter 7 and Appendix G, contaminant migration patterns may be little understood and highly uncertain. The Hanford Groundwater/Vadose Zone Integration Project (U.S. Department of Energy, 1998c) (see Sidebar 4-1 in Chapter 4) may address significant uncertainties and data gaps in the current understanding of the inventory, distribution, and movement of contaminants in order to develop comprehensive risk assessments, with the vadose zone, groundwater, and the Columbia River as receptors, in support of ongoing site cleanup.

- *Future developments in society and technology.* As noted in greater detail in the next chapter, the magnitude of societal or technological changes can be difficult, if not impossible, to anticipate or predict, particularly over the course of decades or centuries. Some such changes can lead to reduced risk, particularly when new developments in science and technology lead to new options for contaminant remediation. Other changes can increase risks by creating new exposure pathways or by bringing increased human populations into areas that were once considered remote. Just 150 years ago, for example, there would have been no concern about drilling into buried waste while exploring for or exploiting natural resources. In addition, U.S. metropolitan regions have roughly doubled in area over the past 25 years, with certain of these, like Denver, now expanding outward toward contaminated DOE facilities at Rocky Flat.

- *Uptake by humans and other species.* Equally uncertain in many cases are the processes by which contaminants travel through and affect exposed organisms. In this regard, controversies continue concerning issues such as linear, no-threshold dose/response models or, in contrast, models based on the concept of hormesis (i.e., the concept that very low doses of toxic substances may sometimes be beneficial) (National Council on Radiation Protection and Measurement, 1995; United Nations Scientific Committee on the Effects of Atomic Radiation, 1993; Jaworowski, 1999). Moreover, the future situations in which humans and other species may be exposed to contaminants also present uncertainties, in part because the behavior patterns of future generations are difficult to predict.

- *Modeling limitations.* As discussed further in Chapter 7 and Appendix G, mathematical models may oversimplify processes, they may use the wrong parameters and relationships among parameters, or they may embody the wrong conceptual structure for the problem at hand. Each of these possibilities creates uncertainty about the accuracy of descriptive or predictive mathematical models.

SCIENTIFIC AND TECHNICAL CAPABILITY

In the context of this report, *technical capability* refers to whether contaminant reduction and isolation measures can achieve site disposition goals—either final, end-state goals, or goals for a desired interim state. *Scientific capability* refers to our ability to understand and conduct the behavior of residual wastes and the environments in which they reside, thereby determining the efficacy of the contaminant reduction and isolation measures being employed, or to know upon which such measures we should rely. Scientific and technical capabilities thus affect the balance among the three "legs of the stool," by affecting the likely effectiveness of the contaminant reduction and isolation legs.

If the technical capability of contamination reduction is good, then cleanup for unrestricted future use may be possible, or if the technical capability of contaminant isolation is good, then controls on site use may figure somewhat less importantly. However, stewardship is likely to remain important because monitoring isolation effectiveness, and intervening if necessary, will have to remain a long-term institutional responsibility, as might additional decontamination of the isolated wastes as technologies capable of doing so become available. In contrast, if the technical capability to achieve either contaminant reduction or contaminant isolation is poor, then stewardship activities become all the more crucial. Theoretical and practical feasibility are important boundary conditions in specifying goals, helping to determine not just whether a goal can be met at all, but the extent to which it can be met (i.e., the extent to which risk reduction can be achieved).

Theoretical Feasibility

The capabilities of technologies have theoretical limits. Thus, it is impossible to separate one substance completely from another. But, in most cases the limits of separations are not important because these limits are far below that which is typically specified as allowable. However, as our contaminant detection ability increases, smaller and smaller contaminant concentrations may cause a technology to fail to meet a remediation goal.

Practical Feasibility

Much more common are limitations on the practical feasibility of contamination reduction or isolation technologies. These limitations, often grounded in basic scientific understanding and technical knowledge, reflect the current status of technology development. For example, it may not currently be possible to locate certain subsurface contaminants, to separate two substances from each other, or to design a barrier that we can assume with confidence will remain intact and compliant with regulations for the thousands, hundreds, or even mere tens of years that may be necessary. Some tasks are simply not possible at this time; others may go part but not all of the way toward meeting a remediation goal. For example, a waste form technology may reduce but not eliminate the migration of tritium or other radionuclides in the subsurface. There is no practical way to separate tritium from groundwater, and in many cases, dense non-aqueous phase liquids (DNAPLS) can not be removed from the subsurface (if, in fact, they can even be detected). At the Hanford Site, there are pump-and-reinject operations around strontium-90 plumes in Area 100 near the Columbia River, but their purpose is to retard migration rather than to remove the contaminants.

In many instances, scientific and technical research and development may eventually overcome limitations in practical technical feasibility if adequate time, expertise, and other resources are available. But in the meantime, limitations on the practical (including costs) as well as theoretical feasibility of technology can constitute a major constraint. The limitation of cost, while often an important factor, is treated separately below. Research and development to improve the feasibility of a technology can also yield lower-cost technologies and methodologies (National Research Council, 1999c).

INSTITUTIONAL CAPABILITY

Institutional capability is, conceptually, parallel to technical capability. It includes considerations about whether the organizations responsible for site remediation and management, the organizations responsible for oversight and enforcement, and other institutions such as the legal system have the ability to carry out their duties effectively over time. As with technical capability, institutional capability affects the balance among the three legs of the committee's metaphorical institutional management stool. In particular, a fundamental question is: "To what extent are institutions able to carry out long-term stewardship activities that can be relied upon as part of the total management system for a residually contaminated site? Realistic estimates of institutional capability are thus an important consideration in establishing interim and end state goals. Institutional capabilities and limitations are discussed more fully in Chapter 7.

There are two important points to stress: the problem of estimating institutional capability, and the lack of adequate framework and adequate empirical data. First, although realistic estimates of institutional capabilities are needed, these estimates are very difficult to make. The ability of institutions to perform stewardship activities reliably over the long term is highly uncertain. This ability has simply not been studied to the same extent as the technical aspects of site disposition, and even with further study, important uncertainties will remain. Second, institutional dynamics, like physical environmental processes, arise from complex interactions among numerous variables, many of which are poorly understood. These complexities may mean that additional data, while helpful, may still not result in the level of understanding that is possible with physical processes. Individuals and institutions may change at rates and in ways that make generalization difficult even when present-day situational aspects of institutional behavior are relatively well understood. Acceptance of a standard model's ability to describe particular phenomena is less common in the social sciences than in the biophysical sciences, and, in many cases, competing models will equally "explain" observed social and institutional phenomena. Past behavior will provide some indication of future behavior, but to date relatively little research funding has been directed toward studies to understand and predict institutional behavior concerning stewardship. Consequently, there are no widely agreed-upon conceptual frameworks for providing assessments of institutions and their stewardship capabilities, nor is there an adequate database for making estimates of future institutional performance.

COST

As used here, "cost" refers to the financial resources and other investments required to transition a waste site from its present state to a desired future state. Included are the costs of contamination reduction and isolation as well as stewardship activities. Cost should be understood not just in terms of money needed by organizations and individuals to perform specific duties or achieve specific ends, but also the "opportunity cost" of then not having the committed resources available for other uses. The latter category can include time volunteered by citizens (e.g., as members of public interest "watchdog" organizations).

Effects of Cost on Site Disposition Decisions

Cost is a key factor (although certainly not the only factor) constraining the current ability to make U.S. Department of Energy (DOE) contaminated sites acceptable for unrestricted use (Probst and Lowe, 2000). For the DOE complex as a whole and for individual site disposition decisions, deciding where and how to spend limited financial resources is a critical contextual factor. At the individual site level, cost typically affects disposition decisions in four ways:

1. Cost concerns at the national level, particularly within the Congress, have had substantial impacts on the pace and timing of cleanup at some DOE sites. They have also led to changes in the way cleanup is being implemented, most notably through recent "privatization" initiatives. At some sites there has been concern that cleanup budgets are now competing with funding for site reuse through private-sector reindustrialization and other community redevelopment initiatives. Whether privatization and reindustrialization will serve to reduce costs (and financial risks) has been a controversial question, in particular the privatization experience with remediation of transuranic waste in Pit 9 at the Idaho National Engineering and Environmental Laboratory (U.S. General Accounting Office, 1997b) and with vitrification of high-level waste at the Hanford Site.

2. Cost is often a consideration—sometimes tacit rather than explicit—in determining the balance among the three sets of measures to achieve risk reduction (contaminant reduction, contaminant isolation, and stewardship). For example, it may be more cost-effective to achieve a specified future state by using a combination of contaminant isolation and stewardship rather than conducting expensive, more complete contaminant reduction measures. However, a future state that includes stewardship is not the same as a future state reached via more complete contaminant remediation, particularly if the latter would allow unrestricted access. The Nevada Test Site, for

example, would be very costly (if at all possible) to remediate, and DOE relies very heavily on the future of NTS as a 'high security' site as a rationale for not cleaning up many areas where the surface and surface environment is contaminated as a result of nuclear testing.

3. At a design level, cost is often an important factor in determining which of alternative techniques should be used to achieve a specific objective. For example, is grouting of waste much less expensive than vitrification, leaving aside other questions such as reliability and effectiveness? Are traditional "hands on" waste exhumation techniques less expensive than using robots, but at the cost of higher risk to the workers? Although computerized records take much less space than paper records and are much more accessible for future analysis by a large group of potentially interested parties, are computerized records less expensive to store and maintain, and what is their lifetime?

4. If the cost of achieving a desired future state is sufficiently large (regardless of the balance among the three sets of measures and of how each measure is designed), the goal of achieving that state may be abandoned at least temporarily and a more modest risk reduction goal may be specified. Alternatively, risk standards may become more lenient or stricter in the future based on a new understanding of risk and effects of dosages on persons and the environment; cost considerations thus may precipitate a tradeoff between future use goals and the stringency of regulations prescribing risk standards. As discussed below, the views of interested and affected parties may affect these tradeoffs.

Cost Considerations

In principle, calculating the monetary cost of a proposed set of site disposition measures is straightforward. One simply specifies which measures will be implemented, determines the amount of material, equipment, land, and labor that is required for each measure, obtains the unit price for the material, etc., and then "does the math." In practice, however, cost estimates (like risk and institutional capability estimates) can have significant uncertainties:

- *Site characterization.* If the site has not been adequately characterized, the actual problem may be very different from the one for which the cost estimate was prepared.
- *Technology.* The contaminant reduction or isolation technologies may be experimental (and thus their costs may be difficult to estimate), their durability (and thus the frequency of incurring additional cost) unknown, or they may not work (requiring further investment to achieve risk reduction goals). The same can be said for stewardship activities. A 1995 DOE internal review of technical and cost assumptions for the Hanford Site tanks program concluded that too many first-of-a-kind technologies were required for remediation of the tank wastes to make realistic cost estimation possible (described in National Research Council, 1996d, pp. 22-23).
- *Duration.* The time over which a site disposition measure will be needed (e.g., institutional controls, "pump-and-treat" technologies) may be uncertain or may have been erroneously estimated.
- *Scope.* The full scope of the disposition effort may be difficult to estimate or may not have been taken into account. (e.g., the cost of off-site disposal of certain wastes may be unknown, the full cost of facility decontamination and decommissioning may have been overlooked, or the characterization of the contaminants in terms of types and amount may be erroneous.)
- *Pricing assumptions.* The emergence of privatization efforts within DOE further complicates cost estimation. Under the DOE standard contracting practices, cost estimates are based on the estimated aggregate costs for the development and deployment of the technologies to be applied. Under privatization, DOE expects to pay the unit costs for the remediation services ultimately provided by private contractors.
- *Predictive economic assumptions.* Assumptions will have to be made about individual price trends, general inflation rates, etc. These assumptions have inherent uncertainties, especially with attempts to forecast costs far into the future.

Despite these uncertainties, reasonably accurate cost estimates can, with some effort, be obtained for many site disposition decisions. In general, however, cost estimates for proven technologies to be applied within the near future are more likely to be accurate than cost estimates of complex, long-term site disposition decisions where the technology that will be applied may still be in the development stage. Costs of the latter still need to be estimated, but the range of such estimates based on uncertainties must be recognized.

Cost Controversies

In addition to controversies arising over how much money should be spent, where, when, and in what ways, controversies can arise over cost estimates. Some conflicts can arise over the calculation methods discussed above. In addition, there are at least three other sources of cost estimate controversy:

- *Discount rates*. In performing calculations about costs to be borne in the future, discount rates often are used to monetize the value of those costs in today's terms. The larger the discount rate, the lower the future cost will appear to be.
- *Hidden costs*. Transaction costs and other hidden costs may be difficult to estimate, yet the experience to date with the Superfund program and the DOE site cleanup program suggests that these costs are often large.
- *Cost shifting*. Costs may also be hidden by "cost shifting," when responsibilities are shifted from one organization to another (e.g., from the federal government to state governments, or from governments to citizen watchdog groups) but are not adequately compensated.

LAWS AND REGULATIONS

As used here, the phrase "laws and regulations" includes the body of civil, criminal, and administrative law at all levels of government, including rulemaking pursuant to these laws, and compliance agreements. The disposition of contaminated sites is addressed at the federal and state level through programs and procedures established under statutes such as the Atomic Energy Act, the National Environmental Policy Act of 1969 (NEPA), the Uranium Mill Tailings Radiation Control Act (UMTRCA), the Resource Conservation and Recovery Act of 1976, as amended (RCRA), the Comprehensive Environmental Response, Compensation, and Liability Act of 1980, as amended (CERCLA), the Federal Facility Compliance Act, the Toxic Substances Control Act, the Clean Air Act, and the Safe Drinking Water Act (see Appendix E). Laws such as these specify goals and methods to be used in making and carrying out site remediation decisions. In addition, other federal and state laws (those concerning budgets and appropriations; property rights, responsibilities, and transfers; torts; contracts; insurance; etc.) provide a legal context within which these decisions take place. Federal facility compliance agreements (as used here, agreements between DOE, U.S. Environmental Protection Agency [EPA], and the state in which a DOE waste site is located) provide further context for these decisions by specifying schedules, budgets, and oversight arrangements to attain particular goals.

Flexibility and Accountability

An ideal legal and regulatory framework would allow flexibility, but require accountability while minimizing conflict. In practice, however, this balance is often difficult to achieve because laws and regulations (although not compliance agreements) are intended to be of general application and cannot anticipate specific situations. Some laws and regulations lean toward stipulating in detail what must or must not be done, while others lean toward establishing general standards and procedures while permitting a good deal of discretionary latitude. For example, the present (1999) statutory and regulatory framework of UMTRCA requires a design-based approach to contaminant isolation. It also requires government ownership of some sites forever. In contrast, the wording under CERCLA expresses a general preference for remediation (treatment of contaminants), addresses what must be done when federal land that has been contaminated is transferred, and acknowledges that institutional controls may be necessary in some situations. The UMTRCA is relatively prescriptive, whereas by comparison CERCLA is more open-ended.

Change

Laws and regulations are always subject to interpretation and change. Formal changes typically occur by statute or through rulemaking; interpretations typically occur through court cases or through guidance documents and policy statements by the regulating or implementing agency. The impetus for change may come from a variety of sources: for example, increased use of a particular remedial approach (e.g., stewardship activities as a prominent component of site remedies); the emergence of new scientific and technical understandings (e.g., the widespread presence of dense non-aqueous phase liquids—DNAPLs—in groundwater with no adequate remedial technology to remove them); or an altered political climate (e.g., receptivity to arguments by responsible parties about the relative costs and benefits of regulatory compliance).

Federal facility compliance agreements are also subject to renegotiation and change. For example, the Hanford Triparty Agreement among DOE, the state of Washington, and EPA calls for a negotiated cleanup schedule. Failure to reach a negotiated schedule results in the opportunity for the state to unilaterally impose a cleanup schedule. One deadline for reaching a negotiated schedule, in this case for the tanks program, came and went with no schedule presented. Rather than imposing its own schedule, the state agreed to give DOE more time to try to negotiate one (*Daily Environment Report*, February 9, 2000, page A-4). Since changes to the legal and regulatory framework are inherently a political process, it is often difficult to predict how the framework will evolve. Compliance agreements are also subject to the political process.

VALUES OF INTERESTED AND AFFECTED PARTIES

As used in this report, ***interested and affected parties*** include individuals or groups that have an interest in site disposition but are not directly responsible for site management or oversight. A discussion of interested and affected parties is found in the report *Understanding Risk* (National Research Council, 1996a).[2] The processes embodied in laws such as NEPA and CERCLA provide opportunities for broad public involvement through public meetings, public hearings, and written comments. In addition, in the mid-1990s DOE initiated the concept of "site-specific advisory boards," which draw representatives from various interested organizations and population subgroups in the area surrounding a DOE facility to provide recommendations on environmental restoration and waste management decisions concerning the facility. Moreover, at many DOE facilities, groups have formed of their own accord to monitor remediation activity and promote their various interests and viewpoints.

Levels at Which Influence is Felt

The views of interested and affected parties can have important effects on how other contextual factors, such as cost and risk, are treated in site disposition decisions. They may influence site disposition decisions in varying directions and strength of influence at five levels of generality:

1. They may help to define risk levels specified in regulations.
2. They may influence priorities about which sites within a facility are addressed first, and to what extent (thereby also influencing the management of other waste sites within the facility).
3. They may help to specify a desired future state for a site, particularly in terms of its preferred future uses.

---

[2] The term "stakeholders," which is sometimes used as an equivalent to "interested and affected parties," is often taken in practice to refer to those with material interests who, by virtue of their jobs as well as their personal well-being, have a stake in site disposition decisions (e.g., site managers and regulators, people living near the site now or in the future). Here, we use the broader and more inclusive term employed in a recent National Research Council report on risk decisions that "... *interested and affected parties* ... may include people from diverse geographic areas, ethnic, or economic groups and organizations.... The parties' concerns may focus on various possible forms of harm, not only mortality and morbidity, but also physical, social, economic, ecological, and moral effects...." (National Research Council, 1996a, p. 87).

4. They may help to decide the relative balance of contaminant reduction, contaminant isolation, and stewardship activities to be used in achieving a desired future state for the site.

5. They may influence choices concerning specific approaches and techniques (e.g., a preference for vitrification over grouting, a desire to have deed restrictions as well as zoning, or an objection to the use of on-site incineration).

Varying Direction and Strength of Influence

Interested and affected parties do not always hold the same views; sometimes, in fact, they may be diametrically opposed. Nevertheless, at a given site and point in time there may be a view that becomes dominant, whether by virtue of its number of proponents, their outspokenness, or their influence over local politics and the local economy. In addition, those with management or oversight responsibilities for a site often live in the community in which the site is located and may, over time, develop close ties with local leaders who are seeking to influence site management decisions. Those responsible for site management or oversight may also change jobs within the community, crossing over to become local leaders and, in some cases, strengthening the dominant view.

In some cases, the dominant view may favor making a site acceptable for unrestricted use, even if funds are scarce and current technical capability is limited. In other cases, however, the dominant view may favor inexpensive remedies and rapid reuse, even if it means restricted use. At the former K-25 area (now the East Tennessee Technology Park) at the Oak Ridge Reservation, buildings are being aggressively marketed for lease by the Community Reuse Organization of East Tennessee (CROET). As an example of the lease arrangements, lndustries leasing space in the building, formerly used for milling and fabrication, are responsible for cleanup of the areas they use, but only to 8 feet off the ground. They are required to keep their operations confined to below that level.

The dominant view may moderate, however, as information is shared among interested and affected parties. For example, many members of the community surrounding the Fernald Site in Ohio originally supported the removal of all contaminants from the site. After extensive fact finding and dedicated participation by interested and affected parties, a site remediation plan was developed and agreed upon that allowed the creation of an on-site waste disposal cell. Such possible changes in the preferences of the public and the makeup of the communities over time must be recognized.

In addition to varying directionality, there are varying degrees of strength in influence. In some instances the input of interested and affected parties has been pivotal to site disposition decisions (e.g., the goals for removing waste from Hanford Reservation tanks, the decision to cap certain waste burial grounds at Oak Ridge Reservation in Tennessee, and the industrial reuse of parts of the Mound Plant in Ohio). In contrast, there are situations where the views of interested and affected parties have seemingly had little effect on site disposition decisions.

OTHER SITES

A number of other sites can influence disposition decisions concerning the waste site in question. These other sites can be categorized as:

- nearby contaminated sites;
- nearby property outside the facility;
- receptor sites; and
- similar sites.

Each is discussed below.

Nearby Contaminated Sites

In many cases, contaminated sites are located within a larger contaminated area. For example, waste burial grounds tend to be built close to each other to take advantage of natural features, to facilitate the burial grounds' operation, and to make security measures easier. In addition, if waste sites have leaked, nearby contaminated soil and water may come to be viewed as a distinct contaminated site. The close juxtaposition of contaminated areas or of contamination problems of qualitatively different types can both complicate remediation planning and limit the ability of cleanup goals to be achieved. Groundwater does not respect site boundaries, a reality that may necessitate a broader context than that of the individual site (or "operable unit") for specifying the desired future state. The implication for long-term stewardship is that the remediation of individual sites may be directed at end uses that, if implemented, would have high probabilities of failure given the larger site context.

Nearby Property Outside the Facility

DOE facilities do not exist in isolation. Each is surrounded by property (land and/or water) that is not under DOE control. To the extent that a waste site is near the facility boundary or has contaminants that may migrate across the boundary, this outside property can affect and be affected by the waste site. Outside property affects disposition decisions because it may present potential for exposure to contaminants. Actions on property outside the waste site (e.g., a more intensive use of a buffer zone or use of resources such as water flowing from the site) may increase the possibility of human exposure to contaminants. For example, sites in the arid western U.S. such as the Nevada Test Site were selected in part on the assumption that nearby population density and water demand would remain low, but the rapid population increase in recent decades in Las Vegas, Nevada, with a consequent expansion of its water demand and settlement boundaries, is clear evidence that this assumption may be wrong. To deal with greater exposure possibilities arising from changes in off-site activities, more elaborate measures (contaminant reduction, contaminant isolation, and/or stewardship) may be necessary on site.

In addition, changes in the type and intensity of surrounding land and water use can affect the physical characteristics of the waste site in question. For example, changes in water use can affect hydrological conditions at the waste site, which can in turn affect the performance of contaminant isolation technologies. At the Hanford Site it has been suggested that irrigated agriculture in areas to the north of the City of Richland could have the beneficial effect of creating a groundwater mound that could help assure protection of groundwater in nearby industrial areas from site-derived contaminants. Similarly, macroscale changes such as global climate change may have unanticipated effects on the waste site.

Receptor Sites

Any remediation activity produces primary wastes (e.g., high-level waste forms and low-level and mixed waste packages) and secondary wastes (e.g., contaminated equipment and fluids, incinerator ash) that must be managed, and contamination reduction by waste removal may generate a large amount of additional waste. Often, the destination of these wastes is another facility (owned either by DOE or a private company), which may be far from the originating site. As a consequence, while risks at the originating site usually are decreased ("usually," because cleanup and transportation worker exposure may entail risks), risks may be increased at the receptor site as well as along transportation routes. Receptor sites can affect disposition decisions at the originating site in a number of ways. Of these, two stand out.

First, the risks may not be acceptable to the receptor site, as well as to those along the transportation routes. For example, the Tennessee state government has taken the position that use of the mixed waste incinerator at the Oak Ridge Reservation is to be restricted to on-site wastes except in "emergency" situations. As another example, the residents of Santa Fe, New Mexico, concerned about the transport of transuranic wastes through Santa Fe to the Waste Isolation Pilot Project (WIPP) site near Carlsbad, New Mexico, successfully initiated a movement to build a bypass.

Second, even if the receptor site does accept the waste, its waste acceptance criteria can shape decisions concerning contaminant reduction processes at the originating site. For example, the calcined high-level tank wastes stored at the Idaho National Engineering and Environmental Laboratory (INEEL) do not meet waste acceptance standards for the proposed repository at Yucca Mountain, Nevada, and must therefore be further processed, at possibly another site (one option would be to ship the wastes to the Hanford Site for vitrification). As

DOE has recognized in recent "system integration" efforts, the complex-wide implications of various disposition decisions—including cost and other resource efficiencies, net risks, and the equitable distribution of risks—need to be considered but are likely to be fraught with controversy.

Similar Sites

By now, the cleanup of contaminated DOE facilities and other sites is becoming a familiar subject. Some precedents have been established for site disposition for considerations such as relative reliance on contaminant reduction, contaminant isolation, and stewardship under particular site conditions. While these precedents are not usually determinative, they often influence site disposition decisions, both as contemplated by DOE and its contractors and as guided by regulators. Following precedents can be an efficient decision-making device; it can minimize having to reenact expensive and time-consuming decision processes on a case-by-case basis, only to end up with the same answer. Remediation activities produce primary wastes (e.g., high-level waste glass logs and low-level waste packages) and secondary wastes (e.g., contaminated equipment and fluids, incinerator ash). A generic example is the classification of radioactive wastes that then leads to a specific disposal technology without much debate (e.g., uranium mill tailings go into piles; low-level waste goes to existing shallow land burial sites).

Rigorous adherence to precedent, however, can result in inappropriate or distinctly sub-optimal decisions. Seemingly similar sites may in fact have important differences that will affect remediation. For example, techniques to remediate sandy soils may work poorly in clayey soils. In addition, continuing to use a well-established technology can preclude the development and deployment of more effective, less expensive technologies (National Research Council, 1999b). Thus, while precedents can expedite site disposition decisions, they need to be used judiciously, to ensure that they are relevant and appropriate.

INTERACTION AMONG CONTEXTUAL FACTORS WITHIN A CLIMATE OF UNCERTAINTY

For purposes of simplicity, each of the seven contextual factors discussed in this chapter—risk, scientific and technical capability, institutional capability, cost, laws and regulations, interested and affected parties, and other sites—has been treated separately. In actuality, however, these factors interact, and they often cannot be neatly distinguished. For example, technical capability questions may arise at both a site to be remediated and at a prospective receptor site, as may issues concerning risk, cost, regulations, and the views of interested and affected parties. In site disposition decisions, then, balancing among the "three legs of the stool" typically is driven by the interaction of and tradeoffs among these contextual factors. In other words, at any stage in the long-term disposition of the waste site, both the types of contaminant reduction, contaminant isolation, and stewardship measures and the extent of reliance on any one set of measures will be affected by the contextual factors discussed in this chapter.

Moreover, as suggested in this chapter, site disposition decisions often are reached in a climate of uncertainty affecting the available choices as well as the contextual factors. Uncertainties can arise concerning the *site at present* (e.g., its characterization, the efficacy of contaminant reduction and contaminant isolation measures); the *site's surrounding physical and social environment at present* (e.g., external exposure pathways, off-site potentially exposed populations and their sensitivity to contaminants); the *site in the future* (e.g., changes in residual contamination over time, changes in the long-term efficacy of contaminant reduction and isolation measures as well as stewardship measures); and the *site's surrounding physical and social environment in the future* (e.g., changes in the surrounding physical environment and its use, leading to on-site changes as well as to changes in off-site human and ecological exposure to contaminants). Many of these uncertainties are exacerbated by technical and institutional limitations. These limitations, and corresponding capabilities, are discussed in the following chapter.

7

Fundamental Limits on Technical and Institutional Capabilities

". . . Policies do not implement themselves" (Weimer and Vining 1999, p. 401). Sites with significant levels of residual contamination will require long-term institutional management, and planning for such management will require realistic thinking. In particular, a realistic understanding is needed of technical and institutional capabilities and limitations and the way those capabilities and limitations may affect the institutional management of U.S. Department of Energy (DOE) waste sites over time.

It is also important to recognize that institutional management decisions and actions take place within a broader setting. While decisions and actions are the most visible component of residually contaminated site management, they must be supported with effective organizational, financial, and legal structures. These structures, in turn, are shaped by contextual factors such as legal and budgetary realities and political and economic pressures, as well as by societal and technological changes that can promote or inhibit the long-term success of institutional management.

Long-term institutional management of DOE's residually contaminated sites can be conceptualized as a system within which planning, decision making, and implementation of all segments of the system must work well for management to operate as anticipated. In Chapters 3, 4, and 5, the contaminant reduction, contaminant isolation, and stewardship measures (the "legs" of the "stool") central to site management decisions and actions were described. In Chapter 6, various contextual factors (the "rungs" of the "stool"), also affecting the disposition of waste sites, were addressed. In the discussion that follows, we delve more deeply into technical and institutional limitations and the societal foundation upon which the entire management system rests.

TECHNICAL CAPABILITIES AND LIMITATIONS

Our collective capability to understand and manage the technical aspects of contaminant reduction and isolation has improved enormously over the past few decades, and there is every reason to think that improvements will continue. Nevertheless, those responsible for managing the investigation and remediation of a contaminated site often must make a remediation decision (or, more realistically, ongoing remediation decisions) in the absence of sufficient scientific and technological knowledge and experience. In addition to the broader challenges of institutional management that will be discussed later in this chapter, this fundamental dilemma is manifest in at least three broad areas:

- inadequate *site characterization*;
- inadequate understanding and monitoring of the **behavior of chemicals and radionuclides in complex environments**; and
- *performance uncertainties* of the candidate technologies.

Site Characterization

Adequate site characterization would result in sufficient knowledge of the site, its contaminants, and the surrounding environment to make an informed site disposition decision. Each aspect of site characterization, however, may be hampered by scientific and technological limitations.

The Site

Quantitative information concerning a residually contaminated site may be sparse or only partially available, in part because of the absence of records of what, where, and how much of hazardous and radioactive materials were disposed of into the environment, particularly during the early operations of the site. The zones of contaminated soil and groundwater and their heterogeneity and extent may be unknown, or known only in a general sense. If data are lacking and conceptual understanding of site dynamics as mediated by physical, chemical, and biological processes is poor, contaminant behavior will similarly be poorly understood. At the Hanford Site, the Idaho National Engineering and Environmental Laboratory (INEEL), and the Nevada Test Site, all sites once thought to possess relatively simple hydrologic and geologic characteristics, contaminant migration has recently been found to be much different from what had been expected (National Research Council, 2000b). Repeated discoveries of such "surprises" regarding the nature of contaminant transport have clear implications for the implementation of long-term institutional management plans at DOE sites.

The Contaminants

In situ waste characterization remains difficult at DOE sites. As one example, consider the Subsurface Disposal Area at INEEL, where DOE is pursuing a pilot characterization approach for buried transuranic (TRU) waste (waste that is contaminated with alpha-emitting transuranium radionuclides with half-lives greater than 20 years and concentrations greater than 100 nCi/g at the time of assay). The focus is on a subsection of Burial Pit 9 of the Subsurface Disposal Area prior to recovery of the contaminants and remediation of the surrounding environment. The characterization approach consists of downhole logging of TRU radionuclide levels, together with sample core collection and analysis. Wastes will then be retrieved to verify the ability of the characterization approach to locate TRU. Plans call for the excavation of these wastes and the processing into acceptable waste forms for disposal in the Waste Isolation Pilot Plant (WIPP) in New Mexico. These efforts have brought out the need for technology development (e.g., remote sensing approaches) to characterize buried wastes.

Still another example is provided by differences in our understanding of acidic versus basic wastes. Most DOE wastes were produced in the operation of acidic separation systems such as PUREX (plutonium and uranium extraction, a solvent extraction process used at the Hanford and the Savannah River Sites); an exception was the metallurgical process used at Rocky Flats, Colorado. Current knowledge of the behavior of plutonium in such acidic wastes is derived from extensive investigations of acidic systems used for a long time in plutonium separations and processing. However, most of the DOE acidic waste in the United States was made highly basic for storage in mild steel tanks to avoid tank corrosion. Unfortunately, the chemical and reduction/oxidation (redox) behavior of plutonium in such basic media is very complicated and much less understood. Even though significant progress has been made in the last decade, more and better data on behavior in neutral and basic solutions is needed before reliable modeling of chemical separations and remediation of soil contaminated via tank leaks and overflows of these wastes can be achieved.

The Site Environment

In many cases, the inhomogeneity of the hydrology and geology of the site and its proximate environment are not sufficiently known to reliably forecast and model potential pathways for contaminant recovery and long-term migration. Pathways may include, for example, continuous and discontinuous fractures in the soils and rocks, permeable sand wedges that cut through strata, and folded and faulted rock strata. Scientific and technological limitations contributing to this characterization problem are discussed immediately below.

Behavior of Chemicals and Radionuclides in Complex Environments

In making decisions about contaminant reduction and isolation technologies, it is essential to understand the behavior of chemicals and radionuclides in complex environments. It is notable that most major DOE contaminated sites are complex due to the geology and hydrology, the waste composition and form, or both; of particular complexity is the unsaturated, or vadose, zone. To achieve the necessary understanding there are essentially two methods: empirical (gathering data and learning from experience) or analytical (use of descriptive or predictive models). The two methods are interrelated; models need data and are based in part on observation, and observation and data are often made intelligible by theory. Below, the capabilities and current limitations of the two methods and their interconnections are briefly discussed.

Learning From Experience

Since the late 1970s our understanding of the behavior of chemicals in complex environments has increased significantly. For example, waste site investigations through the 1970s and into the 1980s failed to recognize and take into account the degree to which the release of liquids with limited water solubility, such as chlorinated solvents (dense non-aqueous phase liquids, or DNAPLS), would complicate and exacerbate both subsurface site characterization and, ultimately, the effectiveness of an aquifer restoration strategy. Consequently, the pump-and-treat approach often was selected in the hope that this technology would restore contaminated groundwater to the desired quality in "reasonable" periods of time (a few years). In actuality, the mass transport limitations posed by DNAPLS, and found to be present to some degree in any heterogeneous subsurface environment, prevent pump-and-treat technologies from removing sources of contamination from many, if not most, contaminated aquifers in time periods less than tens to hundreds of years (National Research Council, 1994c).

This example illustrates on the one hand the limitations of science and technology, and on the other the capability to improve by learning from experience. In some instances, however, learning from experience is simply not possible or the risks are too great. Then especially, there is a temptation to rely on models (which are conceptual or mathematical expressions, simplified to some extent, of how one perceives a system) to complement what can be learned by experience about the behavior of chemicals and radionuclides in complex environments.

Models

As discussed in Chapter 6, risk assessments typically are used to evaluate the hazards posed by sites where contamination will remain, while the term "performance assessment" usually refers to an evaluation of the extent to which an engineered system satisfies its predetermined design or performance criteria. Most performance or risk assessments use **mathematical models**, usually implemented on computers, to describe and predict the fundamental transport and fate processes of both the engineered system and its environment (see Sidebar 7-1). Because of the importance of modeling to much decision making concerning institutional management, the committee has included a discussion of mathematical models in Appendix G.

To be useful, mathematical models rely on the best information about the site, including its physical, chemical, geological, and hydrological properties, and the routes and timing of contaminant exposure to human and environmental receptors. This information determines what parts of the models are deemed relevant in a particular situation and what parameter values and forcing terms (source terms, initial and boundary conditions) are entered.

> **SIDEBAR 7-1**
>
> **ROLE OF MODELS, SITE DATA, AND SCIENCE AND TECHNOLOGY
> IN RISK ASSESSMENT AND MANAGEMENT**
> (by Shlomo P. Neuman and Benjamin Ross; from Appendix G)
>
> - Models are appropriate, often essential, tools for risk assessment and decision making concerning cleanup and management of contaminated, or potentially contaminated, sites. However, it is inappropriate to use models as "black boxes" without tailoring them to site conditions and basing them firmly on site data. Neither disregard of models nor overreliance on them are desirable.
> - The environment constitutes a complex system that can be described neither with perfect accuracy nor with complete certainty. It is imperative that uncertainties in system conceptualization and model parameters and inputs be properly assessed and translated into corresponding uncertainties in risk and decisions concerning risk management. The quantification of uncertainties requires a statistically meaningful amount of good-quality site data. Where sufficient site data are not obtainable, uncertainty must be assessed through a rigorous critical review and sensitivity analyses.
> - Models and their applications must be transparent to avoid hidden assumptions. Model results must not be accepted blindly because hidden assumptions are easily manipulated to achieve desired outcomes.
> - Decisions concerning site disposition and risk management should account explicitly and realistically for lack of information and uncertainty.
> - The monitoring of site conditions and contamination is an imperfect art. It is important that uncertainty associated with monitoring results be assessed *a priori* and factored explicitly into site remedial design and post-closure management.
> - Where uncertainties in science and technology are barriers to effective and appropriate site characterization, remediation, monitoring, and analyses, a suitable research and development program should be initiated and pursued vigorously. The goals of this program should be both short- and long-term. The program should engage a broad array of talents and specialties from government, industry, and academia in order to maintain a proper balance between disciplines and basic as well as applied research.

Nevertheless, it is now widely recognized that the subsurface is a complex, multi-scale, spatially variable natural environment that cannot be fully characterized. Hence, the results of even the most thorough site characterization and monitoring efforts are often ambiguous and uncertain. It is important that models reflect these ambiguities and uncertainties explicitly and, whenever possible, quantitatively.

DOE and other organizations often rely heavily on models for decisions about site remediation and waste disposal. Models have been used to "demonstrate" that a potential waste disposal site or remedial option complies with regulations and is therefore safe, an often fallacious inference (see Sidebar 7-2). Often, models have been used without a serious attempt to validate them against site data. This is especially true of one-dimensional "multimedia" or "multiple-pathway" dose and risk assessment models (such as RESRAD, MMSOILS, MEPAS, and DandD). These models, which are based on a limited menu of highly simplified conceptual frameworks, are used for screening as well as more advanced investigative purposes. They are often used with generic parameters and inputs rather than with site-specific data and are often insufficiently calibrated against actual site conditions. This is also true, albeit to a lesser extent, of more complex two- and three-dimensional subsurface flow and contaminant transport models that incorporate various details of site geology. The tendency to rely on models without detailed site investigations, site monitoring, and field experimentation is sometimes used to justify decisions that additional site or experimental data would be of little value for a project. The reasons for this practice are sometimes identified as regulatory and budgetary pressures.

SIDEBAR 7-2

EVALUATION OF NEVADA TEST SITE GROUNDWATER MODELING
(by Shlomo P. Neuman, member of the DOE/NTS External Expert Peer Review Panel)

In Sidebar 7-1 of this report, models are described as appropriate, often essential tools for risk assessment. The following describes an evaluation of major modeling work conducted at the Nevada Test Site (NTS), work that does not appear to meet many of the conditions listed in Sidebar 7-1. This sidebar was prepared by Shlomo P. Neuman, member of the U.S. Department of Energy (DOE) NTS External Expert Peer Review Panel, from material approved for release by DOE.

Over the past 40 years close to 900 nuclear devices were detonated underground at the NTS as part of the U.S. program of nuclear weapons testing. Many of these devices were detonated at depths near or below the water table so that there is a significant potential for groundwater contamination by radionuclides generated during underground explosions. The DOE Nevada Operations Office (DOE/NV) initiated the Underground Test Area (UGTA) Project to evaluate the effects of underground nuclear weapons tests on groundwater. The Nevada State Division of Environmental Protection regulates the corrective action activities of the UGTA Project through a Federal Facilities Agreement and Consent Order. The individual nuclear test sites have been grouped geographically into six different Corrective Action Units (CAU).

Phase I of the UGTA Project is a Data Analysis Task whose goals include the development of groundwater flow and tritium transport models for the NTS and assessment of risks to human health and the environment, at the regional level. An External Expert Peer Review Panel of scientists was appointed by DOE/NV to examine these modeling and risk assessment efforts. The panel concluded that three-dimensional groundwater flow and contaminant transport modeling of the kind developed under the UGTA Project are appropriate for use in evaluating risks at the regional level within a complex geological setting. The program has gone to considerable lengths to establish a geologic and hydrogeologic model that is reasonably true to data and observations. However, the model does not adequately address the large uncertainties associated with tritium source inventory, the geologic model, controlling flow and transport parameters, and associated risk factors. The panel therefore concluded that the summary statement in the DOE report concerning human health, according to which "... risk to members of the public from subsurface migration of tritium in groundwater is not expected to result in an unacceptable risk as long as human activities involving groundwater remain greater than 10 km from the detonation point during the next 30 years," is not supported by the underlying information presented in the report.

The review panel noted that risk assessment was done only for tritium, while the risk associated with other radionuclides remains unknown. This is so despite the fact that the majority of these radionuclides are more "toxic" than tritium, will persist in the environment for thousands of years due to their long half-lives, and the potential exists for a few of them to migrate almost as rapidly as tritium in groundwater at the NTS. A risk management framework that incorporates spatial dimensions and levels of risk is needed but has not been developed for the NTS.

In the opinion of the review panel, greater emphasis should be placed on modeling uncertainty as a means for determining critical monitoring locations and additional field experiments that are needed to develop reliable observations and predictions at the scale of the CAUs. Wherever possible, model results and predictions should be evaluated against available monitoring data to provide overall weight of evidence for the assessment of risk. There is no indication that such comparisons were attempted. All in all, the panel believes that the project could benefit from a better balance between modeling and data collections efforts, with data collection supporting the modeling and modeling serving to identify where data would be most useful.

The panel looked specifically at the UGTA Project underground nuclear weapons tests in the Frenchman Flat basin, the southernmost CAU at the NTS. It found that, because of data limitations and ineffective modeling strategies, the very limited extent of contaminant migration (a few hundreds of meters) that was

(continued)

> **SIDEBAR 7-2 (Continued)**
>
> predicted to occur in the alluvial aquifer, though possible, has not been established with the degree of confidence that would normally be expected at such contaminated sites. The panel concluded that uncertainty in model predictions was underestimated primarily because alternative geologic and hydrologic conceptual models were not adequately considered in the uncertainty analysis. The models were replete with assumptions that have not been adequately verified by field and laboratory measurements. There were also concerns that the existing data are not adequate to predict the rate of release of radionuclides from test sites or radionuclide reactions with the surrounding rocks. The exclusion from the study of classified radionuclides further increases the uncertainty in model predictions of future radionuclide doses in groundwater.
>
> In the panel's opinion, the current level of problem identification in the Frenchman Flat CAU is not acceptable. Additional field data are needed simply to see whether problems exist or not. Given the current level of information, it is not possible to unequivocally determine the direction of groundwater flow, let alone whether any contaminant plumes have developed in the flow systems at the site. Current model predictions suggest that no such problems exist, but there is almost no field evidence to back up these claims. The panel knows of no precedent where a no-further-action recommendation has been reached at a potentially contaminated site without a much better understanding of the hydrogeological environment and some field confirmation of the model-generated predictions of contaminant distribution.

It may be tempting to use a model to support a decision that a given waste disposal or remedial option is safe, or that additional site data would be of little value, by basing the model on assumptions, parameters, and inputs that favor a predetermined outcome. An example is the assignment of lower permeability in a groundwater flow model than is warranted by available data. Similarly, it may be tempting to "cast the model in a good light" by basing it on a unique system conceptualization and by subjecting it to sensitivity and uncertainty analyses in which parameters and input variables are constrained to vary within narrower ranges than are warranted by the available information. Such practices ultimately detract from the credibility of those who employ them. The use of models is essential, but they need to be untainted by predetermined outcomes. Moreover, they need to be specially designed to the site at hand and supported with adequate, unbiased data. To the extent that they cannot be, their limitations need to be recognized when making remediation and waste disposal decisions.

Technology Performance Uncertainties

There are significant uncertainties regarding the performance of many remediation technologies. These uncertainties can result in selection of a technology that has no clear demonstration of its efficacy in a given environment. The choice may be between two or more currently available technologies, of which one technology does a more complete job but poses a higher risk of failure. Or the choice may be between technologies available today and the prospect of technological improvements in the future. A currently available technology may preclude using a more effective technology later, but waiting for further technology development may defer remedial actions that are needed now.

Use of Current and Future Technologies

In Chapter 4 the example was given of using multiple grout and cement barriers to fix waste remaining in a high-level waste tank after most of the contaminants have been removed. This example illustrates the difficulty of using a technology that stabilizes residual wastes, but may effectively preclude using improved, more complete contaminant removal techniques later on. Similarly, there is serious concern about using more aggressive contaminant tank waste removal technologies currently available, such as oxalic acid solutions, because the aggressive technique may attack the tank shell and induce its partial failure. Yet the grout and cement approach to tank waste stabilization also raises concerns about isolation performance and about waste classification. Thus, the limitations of current technologies may force tradeoffs between, on the one hand, preserving access to the residual contaminants while developing more effective techniques, and on the other, "closing" by stabilizing residual waste that cannot now be safely extracted.

Weighing Uncertainties of Technology Performance

Performance assessments for waste isolation technologies (including both barrier and stabilization technologies) often depend on predicting waste transport. As noted above, however, understanding of the waste site, the contaminants, and the surrounding environment is often too rudimentary in comparison with the modeling accuracy needed to demonstrate compliance with current regulations. Consequently, decisions about waste isolation technologies often must be made under conditions of considerable uncertainty. In the face of these uncertainties, decisions can benefit from an estimate of the health consequences if the technology fails completely or to some degree.

Most regulatory criteria are set at exposure levels that are *acceptable* for licensed or approved activities. Nevertheless, future exposure to chemical and radioactive contaminants may exceed acceptable levels if there was insufficient allowance for uncertainty in the performance assessment. If so, it is important to distinguish whether the exceedances are likely to result directly in grave health consequences for the exposed persons, or whether the exceedances are more likely to go beyond acceptable levels into tolerable levels, that is, levels not likely to result in serious adverse effects. When considering alternate courses for remediation, some defense in depth can be provided if one knows whether failure of a barrier or of an institutional control can lead to radioactive and hazardous chemical exposures that may result in significant health and environmental risk. Sensitivity analyses that explore data and model uncertainties should be used for this purpose. There is a need to study systematically the scientific and technical aspects of contaminant reduction and isolation to reveal the capabilities and limitations with the accuracy and detail necessary to provide for and maintain a focused and relevant program of research and development.

INSTITUTIONAL CAPABILITIES AND LIMITATIONS

At most DOE contaminated sites there is a need to understand not simply how institutional management policies should be formally enunciated, but how they are likely to be implemented over time, and in particular how various factors may cause people to behave or not behave in accordance with official policies. This is sometimes called "forward and backward mapping." The discussion that follows emphasizes that expectations for the fulfillment of institutional management policies should not be unduly optimistic. Instead, it should be recognized that institutional management policies are undergirded by broader organizational, financial, and legal structures that are not static; they can change. Although institutional capabilities and limitations have not received much systematic attention, there is a body of existing social sciences literature on issues of institutional capacity.

Organizational Structures

Much has been published in the academic literature in recent years of the notion of "government failure" (that, as for the case with markets, imperfections that are "built-in" frequently prevent government from realizing hoped-for aspirations and efficiencies). The studies, taken in aggregate, suggest that government is inherently better at some tasks than others. One line of particularly relevant interpretation, for example, is that government may work best when serving as a referee between parties of equal power (government's adjudicatory function), but less well when it takes on the role of partisan "player." Where feasible, such findings should be applied to designing management systems, thereby playing to inherent strengths of the governance system and avoiding its inherent weaknesses. The objectives should be to assign long-lasting problems to long-lasting organizations, and select organizational designs that maximize the chances for effectiveness over time.

Near-Term Factors

Organizations, like people, can differ greatly in their personalities, competencies, and sources of motivation. For example, some organizations have operated nuclear power plants efficiently and safely; others have been less successful. Some organizations make a serious commitment to environmental protection; others simply go through the motions. Each organization often displays a consistent tendency, or **organizational culture**. An organizational culture transcends the characteristics of individual workers; moreover, it typically is resistant to change (see Short and Clarke, 1992; Lawless, 1991). Because of its importance, this point is amplified below.

Organizations tend to develop distinctive ways of viewing the world. In the words of Morgan (1986), "Organization rests in shared systems of meaning." These shared systems can be helpful; they can simplify communication and improve cohesion and task coordination. They have the potential, however, of becoming deeply ingrained. This trait is troublesome when the organizational belief system includes what Clarke (1993) has termed the **disqualification heuristic**—the belief that "it couldn't happen here."

With respect to the DOE defense complex, numerous studies have concluded the organization's culture and belief system contributed to the many health and environmental protection problems that arose as a result of site operations. The primary mission of the complex, nuclear weapons production, does appear to have been executed with great competence, and even DOE's severest critics often emphasize that some shortcomings in protecting human health and the environment may have been due to war-time urgency and subsequent cold-war concerns.[1]

It should be emphasized that DOE is by no means the only organization in which failures of institutions have led to increased risks. For example, the President's Commission on the Accident at Three Mile Island (1979) began its investigation looking for hardware problems that precipitated the 1979 incident at this nuclear power facility, but wound up concluding that the overall problem was one of humans, a problem of what the Commission called a pervasive mind-set, both at the Three Mile Island facility and in the nuclear power industry more broadly, that contributed substantially to the likelihood of accidents. Similarly, the 1986 explosion of the space shuttle *Challenger* has been attributed in large part to the "push" at NASA to get shuttle missions launched on a regular schedule (see, e.g., Vaughan, 1997), and the *Exxon Valdez* oil spill was described by the *Wall Street Journal* as reflecting a pervasive lack of concern by both Exxon and Alyeska with their own risk management plans (McCoy, 1989; for a more detailed assessment, see Clarke, 1993).

In addition, certain predictable tendencies appear to influence many large organizations: in particular, the **bureaucratic attenuation of information flows** and the **diffusion of responsibility**. The bureaucratic attenuation of information flows is a phenomenon wherein concerns expressed by on-the-scene workers are not heard by persons at the top. Among organizational analysts (see especially Vaughan, 1997), such a phenomenon is not necessarily seen as entailing a conscious cover-up. Instead, communication is always an imperfect process, and the more "links" in a communication chain, the more imperfect it is likely to be. This phenomenon sometimes is exacerbated, not counterbalanced, by the aforementioned tendency to develop "shared systems of meaning." Not all kinds of information are equally likely to get through an organizational chain of communication, and bad news is particularly unwelcome.

---

[1] The lack of a "culture of stewardship" has been noted by numerous analysts of past experience at DOE facilities: for example, at Fernald (Sheak and Cianciolo, 1993; Hardert, 1993), Hanford Site (Gerber, 1992; Jones, 1998; U.S. General Accounting Office, 1993, 1996), Pantex (Gusterson, 1992; Mojtabai, 1986), Rocky Flats (Lodwick, 1993), and Savannah River Site (Peach, 1988; Shrader-Frechette, 1993; U.S. General Accounting Office, 1989), as well as across the weapons complex as a whole (Dunlap, Kraft, and Rosa, 1993; Herzik and Mushkatel, 1993; Hooks, 1991; Jacob, 1990; Lawless, 1991; Morone and Woodhouse, 1989; Shrader-Frechette, 1993; Slovic, 1993; U.S. Congress Office of Technology Assessment, 1991; National Research Council, 1995b, 1996e, 1999b; The Washington Advisory Group, 1999).

The diffusion of responsibility phenomenon creates additional problems (Freudenburg, 1992). While the division of labor can enhance efficiency, it can also increase the likelihood that no one will take responsibility for broader or commonly shared problems. Both phenomena can have especially severe consequences for organizations that have been developed to manage advanced and potentially risky technologies, be they power plants, oil tankers, space shuttles, or nuclear weapons complexes (see Perrow, 1984; Sagan, 1994; LaPorte, 1996; Rochlin, 1996). Precisely because the technologies are complex, the organizations that manage them must be large and, consequently, prone to bureaucratic problems.

Longer-Term Factors

Even if an organization begins with a strong commitment to safety, a number of factors can cause this commitment to decline over time. One potentially important factor is ***mission change***; another is the ***atrophy of vigilance***. Over time, most organizations undergo subtle or dramatic mission changes. These changes may occur formally, through official pronouncements and commitments (e.g., the shift from weapons production to cleanup at sites such as Hanford and Rocky Flats), or they may occur informally when workers put energy into priorities that are rewarded while ignoring or giving little attention to other responsibilities that are seemingly less pressing. Since institutional management plans will generally be both repetitive in nature and of substantial duration, they can fall prey to formal and, especially, informal mission change.

The "atrophy of vigilance" (see, e.g., Freudenburg, 1992; Clarke, 1999) is a more subtle but important long-term tendency. To understand this tendency, two of its components—***growing complacency*** and ***predictable cost control concerns***—need to be understood. Growing complacency can be illustrated by the *Exxon Valdez* accident. Although ships coming in and out of the Alyeska pipeline terminal in Valdez had not been immune to problems, over 8,000 tankers had gone in and out of the port over more than a decade without a single catastrophe, that is until 11:59 p.m. on March 23, 1989. It may have been the very success of earlier trips in and out of Prince William Sound that helped create a situation in which a tanker was under the control of a third mate and the Coast Guard personnel on duty were not bothering to monitor even the lower-power radar screens that remained at their disposal after cost-cutting efforts a few years earlier.

This example also illustrates a second component of the atrophy of vigilance: predictable cost control concerns. Not just DOE today, but virtually all institutions, public or private, are likely to face periodic pressures to control costs. The sources of pressure may include responses to cost overruns, calls to "cut down on waste and inefficiency," private-sector competition, or simply a desire to do more with less. Whatever the original pressure source and the nature of the organization, at least one response is likely to be consistent: organizations will seek to protect what they regard as their core functions and will cut back on those they regard as peripheral.

Unfortunately, safety measures such as long-term monitoring may be regarded as peripheral or "non-productive," especially if there has been no demonstrated need for them. For example, planned installation of a larger permanent cap over the currently buried highly radioactive residues stored at the former DOE Niagara Falls Storage Site in Lewiston, NY, would have resulted in the loss of 13 inner perimeter sampling locations. The consequence of such an action would be to increase migration distance (and thus, time) before contaminants leaking in the groundwater from the containment structure would be detected, providing an increased risk to nearby residents and public facilities (National Research Council, 1995a).

Legal Structures

A key feature of the legal structure governing the cleanup of DOE's waste sites concerns the U.S. Constitution; there may be no such thing as a "binding" Congressional or federal commitment. Just as the U.S. Congress makes laws, it can unmake them. While its powers are not unchecked (by, e.g., the threat of political opposition, public outrage, or a presidential veto), the ability of Congress to reverse itself always remains. This power gives *flexibility* to undo laws that, in retrospect, were ill-advised or are no longer appropriate; however, it also means that, as noted in Chapter 6, even ***legally mandated arrangements are subject to change***. Thus, if Congress passed a law requiring systematic stewardship for DOE's waste sites, for example, it subsequently could amend or repeal the law or simply fail to appropriate funding to carry it out.

Proposed changes to an existing law are sometimes highly visible and well publicized, but they sometimes occur with relatively little congressional debate, simply by inserting an amendment into an apparently unrelated measure or into a final conference committee action on an appropriation. For example, within less than 10 years of the passage of the much debated and carefully crafted compromises that became the 1982 Nuclear Waste Policy Act, there were at least two major congressional amendments. One amendment, added to the Senate version of the Omnibus Budget Reconciliation Act of 1987, abandoned the specified site selection process for a second high-level and commercial spent nuclear fuel waste repository and established an official preference that the site for the first repository be located at Yucca Mountain, Nevada (Appendix E). The other amendment terminated the guarantee that Nevada could receive independent federal funding for its own scientific research on the suitability of the Yucca Mountain site and on the prospective impacts of the repository siting (Wald, 1992). Both of these amendments were considered quite important to citizens of the affected state, but each was inserted into the final legislation at the "last minute" and passed with relatively little congressional debate (Easterling and Kunreuther, 1995). Such examples illustrate the need for the forthright recognition of the fragility of federal assurances, not just over the course of several centuries or decades, but also over a few years.

Congressional actions can have unanticipated consequences for agencies as well. For more than 20 years, DOE had responsibility for cleaning up sites that were within the Formerly Utilized Sites Remedial Action Program (FUSRAP). Under Public Law 105-62, however, Congress transferred the FUSRAP sites from DOE to the U.S. Army Corps of Engineers, beginning in FY 1998. The subsequent agreement between DOE and the Corps to effectuate the law put DOE in the position of reinheriting the FUSRAP sites two years after the Corps completed remediation. DOE now has only limited input into site remediation decisions but remains responsible for providing long-term institutional management of FUSRAP sites that remain residually contaminated following their remediation. This obligation, already difficult, becomes especially challenging when institutional controls must be specified in detail in decision documents for remedies that include institutional controls (see Chapter 5, Periodic Reevaluation of the Site Protective System).

Financial Structures

Even if policies aren't changed dramatically, carrying out a long-term governmental commitment requires predictable funding. Nevertheless, unless funds are provided in advance, continued funding depends on continued congressional actions to authorize, appropriate, and otherwise see to the actual spending of the "promised" funds. In general, the traditional response to the threat of congressional reversal on funding is to rely on political pressure. Such techniques have been fairly successful to date in obtaining reasonably steady funding for site remediation, especially in politically influential states, but these techniques cannot be counted on over the longer term (and sometimes over the short term), and they thus do not provide a good basis for planning for ongoing institutional management. No matter how genuine a given agency's or official's intentions may be, governmental assurances of future funding are justifiably met with skepticism.

BROAD SOCIETAL FACTORS

Beyond site management activities themselves and their underlying organizational, legal, and financial structures (including reduced spending to lower taxes), broad societal factors also can exert important influences. Three categories of factors are particularly worthy of attention: (a) *"beneficial reuse" pressures*, (b) *societal/technological changes*, and (c) **trust and credibility challenges**.

"Beneficial Reuse" Pressures

Some DOE sites (land and/or facilities) are or will be attractive for economic, residential, or recreational purposes. Recent examples include the transfer of the Pinellas Site, Florida, to the Pinellas County Industry

---

SIDEBAR 7-3

REINDUSTRIALIZATION OF THE MOUND SITE
(by Raymond G. Wymer)

In 1989, the U.S. Department of Energy (DOE) decided to close the Mound Site, changing its mission from support of weapons production to environmental restoration. The Mound Site was added to the National Priorities List (NPL) and entered into a Federal Facilities Agreement (FFA) with the state of Ohio and the U.S. Environmental Protection Agency (EPA). Cleanup activities are conducted under the Comprehensive Environmental Response, Compensation, and Liability Act of 1980, as amended (CERCLA). DOE is working to exit the site by 2005 and to transfer the site to Miamisburg Mound Community Improvement Corporation (MMCIC), a non-profit corporation to coordinate the transfer and economic development of the property. EPA must approve such a transfer under CERCLA section 120(h), whereas the Ohio Environmental Protection Agency (OEPA) has regulatory authority because it is party to the site's FFA. To help in achieving its goals the Mound Site has formed a team of decision makers that includes the DOE decontamination and decommissioning manager, EPA, OEPA, and the Ohio Department of Health. In collaboration with its regulators the site has developed a process to facilitate release of facilities with radiological process histories. This process incorporates generic release criteria established in DOE Order 5400.5 for the release of an intact facility with surface contamination and draft criteria for the release of debris with residual radioactive contamination based on a dose limit in Draft NUREG 1500. Mound Site and its regulators determined that the criteria established in DOE Order 5400.5 are appropriate for release of intact facilities for conditional use. Facilities intact when the Mound Site is transferred must be in accord with deed restrictions for the site, and consequently will be used only for industrial use. Mound Site is currently working with its regulators to identify a dose limit that is acceptable to decision-making authorities. DOE Order 5400.5 and 10 CFR 20 establish a dose of 100 mrem/yr for all exposure modes from all DOE sources of radiation. However, the regulations do not establish a method for apportioning the 100 mrem dose when more than one source is present. Mound Site and its regulators are considering a dose limit of 15 mrem/yr above background.

REFERENCE

U.S. Department of Energy. 1999 (July). A Monograph: Facility Disposition Lessons Learned from the Mound Site. Office of Environmental Policy and Assistance RCRA/CERCLA Division (EH-413) DOE/EH-413-9909, Washington, D.C.

Council, the sale of portions of the Mound Site, Ohio, to the Miamisburg Mound Community Improvement Corporation (see Sidebar 7-3), and the reindustrialization effort at the former K-25 site (now the East Tennessee Technology Park) at Oak Ridge Reservation, Tennessee. In addition, the potential for urban development pressures exists at sites such as Rocky Flats, Colorado, and the Nevada Test Site, both of which were once considered remote but are now experiencing rapid urban growth nearby.

The goal of beneficial reuse of otherwise under-used federal sites is laudable in many ways, but it can pose risks. The Hall Amendment (Section 31544 of the National Defense Authorization Act of 1994 amending Section 646 of the U.S. Department of Energy Organization Act), aimed at promoting that goal, allows DOE to lease its temporarily unneeded or excess acquired real property at closing or reconfigured weapons production facilities. Leases are for periods of up to 10 years, but they can be renewed for more than 10 years if the Secretary of Energy determines that renewal promotes national security or is in the public interest. Before leasing, the DOE is to consult with the EPA for sites listed on the National Priorities List (NPL), or the appropriate state official for sites not on the NPL, to "determine whether the environmental conditions of the property are such that leasing the property, and the terms and conditions of the lease agreement, are consistent with safety and the protection of public health and the environment" (Department of Energy Organization Act, section 646, 42 U.S.C 7256).

Many of the proposed uses entail potential human exposure to contaminants that either remain on site or are migrating from nearby, still-contaminated areas. As discussed in Chapters 4 and 5, the ability to restrict uses under proprietary and governmental controls is questionable, especially over the long term, as is the long-term ability to maintain contaminant isolation barriers and monitoring systems. Moreover, as a day-to-day DOE presence is replaced by contractors, subcontractors, lessees, sublessees, etc., the careful supervision of activities that could result in exposure to residual contaminants is likely to diminish.

Nevertheless, state and local governments often face intense pressures to maximize jobs and development. Authors such as Krannich and Luloff (1991) have noted that in rural areas, leaders often desperately encourage development at virtually any cost (see also Freudenburg, 1991); the same could be said of depressed urban areas. Meanwhile, at the outer fringes of urban areas the pressures to minimize obstacles to economic development are so well known that urban scholars often refer to cities as "growth machines" (see Molotch, 1976). While local residents may have a range of views, often the most powerful and influential local advocates tend to be strongly in favor of the intensification of land use and seek to attract new economic activities, rather than seeing it go elsewhere (see Block, 1987; Edelman, 1964; Logan and Molotch, 1987; Stone, 1989). These advocates are often capable of exerting quiet, behind-the-scenes development pressure long after most members of the public have lost interest and long after records of residual contamination have become lost or forgotten. These pressures can make it difficult for governments to restrain development and to conduct vigorous oversight of residually contaminated sites. While development pressures are not necessarily suspect, they do need to be anticipated as having the potential to undercut present-day intentions.

Societal/Technological Change

One of the few general predictions about the future that can be made with confidence is that society and the available technology are likely to undergo changes that are difficult, if not impossible, to envision in advance. At present, for example, development pressures in Henderson, Nevada, a Las Vegas suburb that is now the state's second largest city, are requiring urgent remediation measures for mines that were abandoned mere decades ago. Nobody anticipated the recent population growth for the Las Vegas region and that humanity would move toward old mining areas (*Craig Daily Press*, December 25, 1999, p. 2). Another example of this urban sprawl is the encroachment of suburban Denver, Colorado, where population has increased five-fold since 1930, toward the contaminated DOE Rocky Flats Site (U.S. Department of Energy, 1999).

Given the rapid increase in suburban sprawl in the latter half of the twentieth century, residential development has begun to press in on many areas that were once considered remote. Indeed, the rates of sprawl have increased significantly in the past two decades, suggesting that conflicts with what are now considered to be remote locations may become problematic sooner than expected. Yet, as has been noted by Erikson, Colglazier, and White (1994), our ability to "see" far into the future may be similar to trying to understand a vast cavern from what can be seen through a small peephole. Realistically, perhaps all that can be done is to plan for the kinds of societal and technological futures that we are able to anticipate. But, when doing so, we need to be realistic about the limitations of our foresight. If the changes of the next century are as great as those of the century just ended, the implications could be quite dramatic. Just 100 years ago, for example, uranium was defined as a mineral with few uses (porcelain glaze, for one); as recently as 150 years ago no one had thought of drilling for oil or, for that matter, drilling more than a few dozens of feet in search of any mineral resource.

Trust and Credibility Challenges

As noted by a wide range of analysts (see, e.g., Dunlap, Kraft, and Rosa, 1993; Jacob, 1990; LaPorte and Metlay, 1996; Slovic, 1991; Secretary of Energy Advisory Board Task Force on Radioactive Waste Management, 1993), one of the central challenges facing the management of DOE waste sites is the legacy of distrust. It is now well known that governmental organizations in general have suffered a decline of deference by the public in the past three decades. Trust and credibility are subject to "the asymmetry principle:" they are hard to gain but easy to lose (on this point, see especially Slovic, 1991, 1993.) At many DOE contaminated sites, past mistakes have led to a severe erosion of trust and credibility. As Rosa and Clark (1999, p. 22) have noted, such nuclear enterprises provide "a paradigmatic example, capturing the essential features of technological gridlock . . . producing a polarization between citizens, on the one hand, and policymakers, experts and managers, on the other hand, with the net result being impasse over technological choices."

Trust and credibility issues have particularly important implications for the long-term management of DOE contaminated sites. Some of these sites are large and remote enough to be good locations for undertaking dangerous experiments. For example, approaches could be tested for cleaning up intractable non-nuclear contamination such as DNAPLs, but for such arrangements to be put into place, DOE or other responsible parties would need to reach understandings with regulators and other interested and affected parties (National Research Council, 1999b). While such understandings are not out of the question, they would be far easier to reach within a context of trust and credibility.

STRENGTHENING LINKS BETWEEN TECHNICAL AND INSTITUTIONAL CAPABILITIES

As is evident from the foregoing discussion, disposition of DOE's waste sites is hampered by difficult technical and institutional limitations. Nevertheless, corrections can be made if these limitations are acknowledged and if links between technical and institutional capabilities are strengthened. Two areas where technical and institutional capability can be mutually reinforcing include (1) periodic reevaluations of site disposition decisions, and (2) the development of new science and technologies.

Periodic Reevaluations

As noted in Chapter 5, a comprehensive approach to long-term institutional management of residually contaminated sites includes periodic reevaluations. To date, there is insufficient evidence to predict whether these evaluations will be meaningful. Based on the EPA record of completing five-year reviews under the Comprehensive Environmental Response, Compensation, and Liability Act of 1980, as amended (CERCLA), there is reason for skepticism about whether periodic reevaluations will actually be conducted when needed, and in a thorough and effective manner. Nevertheless, periodic reevaluations offer an opportunity to reassess how well the total site disposition system, including its technical as well as its stewardship components, are working together to ensure an acceptable level of risk. Through periodic reevaluations, some of the negative impacts of the technical and institutional limitations discussed in this chapter can be reduced, even if they cannot be eliminated.

This chapter has provided examples of cases where vigilance or constancy have been degraded over periods as short as a decade or less, raising serious concerns about the ability of measures like diligent periodic reevaluations to persist for periods of 100 years or more. In part, however, the failures that have occurred have come about because initial policy planning failed to recognize or take into consideration some of the predictable ways the proposed implementation could be disrupted by parties having the incentive and ability to prevent full implementation. In addition, there have been at least some approaches to institutional management that have proven more successful that others, particularly over the short-to-medium term of between 1 and 100 years. Successful approaches appear to have been characterized by incentive structures that appear to be well suited for the types of performance needed. Such incentives may include adequate, stable resources for monitoring and maintenance of contamination, provisions for broad, effective oversight by the public, and establishment of appropriate public use for the area (e.g., nature reserve or park). For the future, accordingly, while there would clearly appear to be value in minimizing the need to rely on fallible human institutions where possible, there may prove to be considerable value in examining more systematically the types of institutional management measures that have proved to be somewhat more successful over the short to medium term.

The relatively high likelihood that institutional management measures will fail at some point underscores the need to assure that decisions made in the near term are based on the best available science. Where deficiencies in

SIDEBAR 7-4

BASIC RESEARCH NEEDS IN SUBSURFACE SCIENCE

A recent report issued by the National Research Council (2000) concluded that basic research is needed in four areas of subsurface science: location and characterization of subsurface contaminants and characterization of the subsurface, conceptual modeling, containment and stabilization, and monitoring and validation. These recommendations are germane to the issues of long-term institutional management of contaminated sites.

Location and Characterization of Subsurface Contaminants and Characterization of the Subsurface
- Improved capabilities for characterizing the physical, chemical, and biological properties of the subsurface.
- Improved capabilities for characterizing physical, chemical, and biological heterogeneity, especially at the scales that control contaminant fate and transport behavior. Approaches that allow the identification and measurement of the heterogeneity features that control contaminant fate and transport to be obtained directly (i.e., without having to perform a detailed characterization of the subsurface) are especially needed.
- Improved capabilities for measuring contaminant migration and system properties that control contaminant movement.
- Methods to integrate data collected at different spatial and temporal scales to better estimate contaminant and subsurface properties and processes.
- Methods to integrate such data into conceptual models.

Conceptual Modeling
- New observational and experimental approaches and tools for developing conceptual models that apply to complex subsurface environments, including such phenomena as colloidal transport and biologic activity.
- New approaches for incorporating geological, hydrological, chemical, and biological subsurface heterogeneity into conceptual model formulations at scales that dominate flow and transport behavior.
- Development of coupled-process models through experimental studies at variable scales and complexities that account for the interacting physical, chemical, and biological processes that govern contaminant fate and transport behavior.
- Methods to integrate process knowledge from small-scale tests and observations into model formulations, including methods for incorporating qualitative geological information from surface and near-surface observations into conceptual model formulations.
- Methods to measure and predict the scale dependency of parameter values.
- Approaches for establishing bounds on the accuracy of parameters and conceptual model estimates from field and experimental data.

The research needs outlined above call for more hypothesis-driven experimental approaches that address how to integrate the understanding of system behavior.

Containment and Stabilization
- The mechanisms and kinetics of chemically and biologically mediated reactions that can be applied to new stabilization and containment approaches (e.g., reactions that can extend the use of reactive barriers to a greater range of contaminant types found at DOE sites) or that can be used to understand the long-term reversibility of chemical and biological stabilization methods.
- The physical, chemical, and biological reactions that occur among contaminants (metals, radionuclides, and organics), soils, and barrier components so that more compatible and durable materials for containment and stabilization systems can be developed.
- The fluid transport behavior in conventional barrier systems; for example, understanding water infiltra-

tion into layered systems, including infiltration under partially saturated conditions and under the influences of capillary, chemical, electrical, and thermal gradients can be used to support the design of more effective infiltration barrier systems.
- The development of methods for assessing the long-term durability of containment and stabilization systems.

Monitoring and Validation

Many of the research opportunities for monitoring and validation have been covered in the research emphases discussed above. In addition:
- Development of methods for designing monitoring systems to detect both current conditions and changes in system behaviors. These methods may involve the application of conceptual, mathematical, and statistical models to determine the types and locations of observation systems and prediction of the spatial and temporal resolutions at which observations need to be made.
- Development of validation processes. The research questions include (1) understanding what a representation of system behavior means and how to judge when a model provides an accurate representation of a system behavior—the model may give the right answers for the wrong reasons and thus may not be a good predictive tool; and (2) how to validate the future performance of the model or system behavior based on present-day measurements.
- Data for model validation. Determining the key measurements that are required to validate models and system behaviors, the spatial and temporal resolutions at which such measurements must be obtained, and the extent to which surrogate data (e.g., data from lab-scale testing facilities) can be used in validation efforts.
- Research to support the development of methods to monitor fluid and gaseous fluxes through the unsaturated zone, and for differentiating diurnal and seasonal changes from longer-term secular changes. These methods may involve both direct (e.g., in situ sensors) and indirect (e.g., using plants and animals) measurements over long time periods, particularly for harsh chemical environments characteristic of some DOE sites. This research should support the development of both the physical instrumentation and measurement techniques. The latter includes measurement strategies and data analysis (including statistical) approaches.

REFERENCE

National Research Council. 2000. Research Needs in Subsurface Science: U.S. Department of Energy's Environmental Management Science Program. Committee on Subsurface Contamination at DOE Complex Sites, National Academy Press, Washington, D.C. 159 pp.

scientific understanding that inhibit present-day planning are recognized, incorporating strategies for improving the scientific and technical basis for future decisions increases the chances that those decisions will be soundly based. At the same time, the deficiencies in institutional performance pointed to in this chapter can work against the long-term interests of research and development as well, and have done so in the past (National Research Council, 1996c). At some level, the same fundamental limits affect both the scientific-technical and the institutional and organizational systems. For this reason, attention to the needs of both is necessary in the design and implementation of institutional management systems, a point pursued in the next chapter.

New Science and Technology Development

The prospect of advances in contaminant reduction and isolation technologies gives reason for optimism, albeit not for complacency. There is a good possibility that future scientific and technological developments will create new capabilities that can, in fact, improve waste and environmental characterization and monitoring. They can also resolve uncertainty and spur contamination reduction and isolation and stewardship that achieve the goal of protecting the public and the environment in the most efficient manner. Scientific and technological developments that are actively sought, however, have greater likelihood of being realized than are those that are merely hoped for. As noted in Chapter 5, science and technology developments can be sought either by monitoring the emergence of new remedial technologies in non-DOE settings (e.g., in the private sector or in other nations) or by direct sponsorship of new science and technology research and development (R&D), through peer-reviewed processes and rational, needs-based selection of R&D projects (National Research Council, 1996e; 1998c; 1999b,c). New science and technology developments entail an institutional commitment that is complementary to periodic reevaluations. Sidebar 7-4 gives some of the recommendations for basic research needs in the subsurface sciences from a report by the National Research Council (2000b). A reevaluation may suggest the need for further contaminant reduction or isolation; new sciences and technology development may suggest a way this can be accomplished. Moreover, both periodic reevaluations and new science and technology developments take advantage of present institutional capabilities to "buy time" to find longer-lasting solutions.

8

Design Principles and Criteria for an Effective Long-Term Institutional Management System: Findings and Recommendations

The foregoing chapters have demonstrated several key points:

1. Only a small number of U.S. Department of Energy (DOE) sites can now be remediated to a level that permits unrestricted use. For most of the remaining sites, only an intermediate level of cleanup and safety is currently possible, enough to permit some uses, but not enough to permit unrestricted use.

2. Everything that will need to be considered in making decisions for the long-term disposition and management of residually contaminated sites cannot be precisely known. Instead, decisions often will have to be made under conditions of irreducible uncertainty.

3. Sustained vigilance will be required, yet there is reason to be skeptical of our collective societal ability to sustain such vigilance.

To return to the prescient observation of Alvin Weinberg noted at the beginning of this report, management of nuclear wastes and contaminated sites will require "both a vigilance and a longevity of our social institutions that we are quite unaccustomed to." Given the scope of the challenges and current limits in scientific understanding and technical capability, these problems cannot simply be made to "go away" by application of existing scientific and technical know-how. Nor is it prudent to adopt the position that the influences that can erode management systems and organizations will be successfully held at bay over time by existing institutions, including but not limited to DOE. Under the circumstances, there are no existing formulas or "cookbook" solutions that can simply be pulled off the shelf and put into place. Instead, the best guidance that can be offered involves two components: first, the need for sober recognition of the magnitude of the challenge that needs to be faced, and second, the need to favor institutional and technical systems that will have high probabilities of being able to anticipate and correct problems, and more broadly, to minimize future regrets.

DESIGN PRINCIPLES AND CRITERIA

The challenge is to do the best one can in dealing with a problem that cannot be fully known in advance. This appears to come down to placing emphasis on prudence and precaution. Specifically, there are nine characteristics of institutional design that need to be emphasized:

- *defense in depth;*
- *complementarity and consistency;*
- *foresight;*
- *accountability;*
- *transparency/visibility;*
- *feasibility;*
- *stability through time;*
- *iteration;* and
- *follow-through and flexibility.*

Defense in Depth: Layering and Redundancy

In Chapter 5, the characteristics of layering and redundancy were discussed. "Layering," as used in this study, refers to using more than one element to accomplish basically the same purpose; "redundancy" refers to having more than one organization responsible for basically the same task. While the concept was discussed in Chapter 5 in terms of stewardship activities, it applies more broadly to the total system of institutional management for the site as well. For example, layering could occur by having both contaminant isolation measures (such as engineered barriers) and stewardship measures (such as deed restrictions and zoning), all intended to accomplish the same purpose of preventing undue exposure to residual contaminants. Similarly, organizations such as the site owner, local citizens groups, and regulatory agencies might all have the right and responsibility to oversee the ongoing management of a site and to ensure that safety measures are performed. Redundancy requires careful coordination and mutual trust to avoid chaos.

Complementarity and Consistency

"Complementarity and consistency" as the phrase is used here, refers to having contaminant reduction, contaminant isolation, and stewardship measures that support and enhance each other, rather than hobbling or detracting from each other. If, for example, a contaminant isolation measure (such as waste entombment) makes it difficult to carry out a stewardship activity (such as monitoring and oversight), the components of the site's institutional management system will not be complementary and well integrated. It is also particularly important to guard against cases where the day-to-day incentives for agency management personnel and contractors run counter to official agency policy. For example, if official policy is to secure the highest possible levels of public safety, it is important to have the agency's official reward system (e.g., salary increases, promotions, contractor bonuses) reinforce that policy rather than being based on other factors that are politically popular or easy to quantify (e.g., number of acres or sites transferred to private hands).

Foresight

"Foresight" refers to anticipating how the components of the system will work, individually and together, and making preparations in a timely fashion. The committee has observed in the past the tendency of DOE to make commitments in such documents as Tri-Party Agreements to remediation actions that are not technically feasible, often resulting in delays and loss of trust in DOE. For example, if a contaminant isolation measure such as a cap requires monitoring, but the monitoring capability is not designed into the barrier system, retrofitting may be difficult and expensive. On the other hand, a barrier system of multiple layers, coupled with monitoring equipment that could detect a failure in the first line of defense before the second is threatened, might reflect better foresight and reduce long-term costs in accordance with the old adage that "an ounce of prevention is worth a pound of cure." Similarly, preparations for legal use restrictions and appropriate systems of enforcement need to be made well before property transfers are conducted. Foresight sometimes may be constrained by uncertainty, but it should be employed to the greatest extent possible.

Accountability: Ability to be Monitored and Enforced

"Accountability" means both "answerable" and "capable of being explained." As used here, it refers to the ability of both the human and the technical components of the site's management system to be monitored and, if necessary, corrected through renewed remediation activities, enforcement, or other means. If people responsible for various site management and oversight tasks cannot be held answerable to the interested and affected public for their actions or non-actions, or if the site's remedial technologies and its physical environment do not perform as expected, yet that deviation goes unnoticed and unexplained, the efficacy of the site's protective system is likely to erode over time.

Transparency/Visibility

"Transparency/visibility" as used here refers to having site disposition decisions that are not only rational, but also clearly articulated and readily accessible to public scrutiny. People need to understand both the site disposition decision and its rationale to be able to evaluate effectively whether the site's protective system is working as anticipated. Without this transparency, the public still may be able to evaluate whether the system is failing, but only after the failure has become evident or more serious.

One of the characteristics of what appear to be relatively reliable organizations seems to be a high degree of openness and visibility such that errors can readily be seen and understood to be problematic by a wide range of people, and those people are reasonably free from what Martin (1999) terms "the suppression of dissent." Under such circumstances there can be a significant increase in the probability that an error will be detected and corrected. This approach runs counter to the tendency to favor organizational secrecy or to solve problems by putting them "out of sight, out of mind." Not only are transparency and visibility needed for an open analytic-deliberative process involving citizens as well as regulators and management personnel (National Research Council, 1996a), but transparency and visibility can improve system safety and lay the groundwork for accountability.

Feasibility

"Feasibility" refers to having an institutional management system that is technically, economically, and institutionally possible to implement within a specified time period. If, for example, the disposition decision calls for a remedial technology that has not yet been fully developed or tested, the system will not be feasible unless this limitation is overcome. Similarly, unwarranted institutional expectations (e.g., expecting local governments to be impervious to development pressures, or expecting DOE to continue to receive high levels of funding for oversight into the indefinite future [see Probst and Lowe, 2000]) may lead to infeasible assumptions about site management.

Stability Through Time

"Stability" refers to the likelihood, based on reasonable estimates, that the components of the site management system and the system as whole will continue to perform as expected. A continued, stable investment in resources must be assumed to accomplish this stability. Stability may be much more likely with some elements, specifically those requiring a minimum of upkeep, monitoring, oversight, and enforcement. In some cases, measures that increase stability may lead to decreased flexibility, particularly in terms of institutional performance. However, analysts such as LaPorte and Keller (1996) have argued that nuclear waste management appears to create greater need for "institutional constancy" than is possible from the typical approach in many policy institutions, namely "muddling through."

Iteration: Revisiting Site Disposition Decisions

"Iteration" refers to the concept that, when a site's uses must be restricted because of residual contaminants, it is desirable to periodically reconsider both how well the site's protective system is working and whether it can be improved. Iterability is thus motivated by both caution (i.e., a recognition of the need to plan for fallibility) and optimism (i.e., a recognition that better technologies or institutional arrangements may become available in the future). While pressure for expanded use of the site and nearby resources (e.g., water) may be the greatest impetus to revisit site disposition decisions, iterability should be routinized. In other words, it should be integrated into the total site disposition decision process.

Follow-Through and Flexibility

Given the impossibility of establishing a "best" solution in advance for the vast majority of DOE sites, it is obvious that there will also be a need for more systematic iteration or reconsideration on a periodic basis. The need for iteration coupled closely with follow-through mechanisms is thus a matter of both caution (i.e., recognition of the need to plan for fallibility) and of optimism (i.e., a recognition that better science and technologies or institutional arrangements may become available in the future, particularly if institutions of the present and near future make the needed investment in research and development). The challenge is that, even though experience to date has provided some evidence of success, there is little evidence that present-day institutions can be counted on to provide follow-through to act on new information with the degree of long-term reliability and rigor that current contamination problems require. Under the circumstances, perhaps the best that can be done is to plan more carefully about the kinds of approaches and institutions that could be developed to do a better job, as noted earlier, keeping in mind at all times the importance and preferability of minimizing regret. By this, we mean developing a decision strategy overall that avoids foreclosing future options where sensible, takes contingencies into account wherever possible, and takes seriously the prospects that failures of institutional controls or other stewardship measures in the future could have ramifications that a good steward would want to avoid triggering through inappropriate action in the present. In practice, as noted earlier, this may mean deciding not to tear down certain structures that may prove to have historical value as future museums or interpretative sites, or choosing to leave sites in federal control as environmental preserves or wildlife refuges rather that starting a process of private commercial development that could well prove to be self-intensifying and difficult to control or redirect. There is also a need to recognize, first, that some options will entail limiting future options but may still, on balance, be preferred; and second, that some options have already been foreclosed by past actions.

In summary, a "one-size-fits-all" formula for institutional management is not advocated in this study. Instead, institutional management should be tailored to the needs and conditions of each site. Nevertheless, the general precepts spelled out in this chapter—the system design criteria, iterative decision process, and requisite attributes of institutional mechanisms—should be considered whenever a DOE waste site will, after remediation, have residual contamination necessitating restrictions on the uses of the site or nearby resources.

The following discussion pertains generally to sites across the U.S. Department of Energy (DOE) legacy waste complex, rather than addressing particular sites and facilities. In addition to information and perceptions gained in the course of this study by presentations from DOE personnel, contractors, representatives of other agencies, other experts, and interested public citizens, as well as from tours of a number of sites, the members of the committee also drew on a wide spectrum of individual background experience and knowledge of many other relevant activities and studies, in particular, current studies by the National Research Council.

FINDINGS

1. Almost All Sites Will Require Future Oversight. Although considerable progress has been made over the past decade in treating and stabilizing wastes at DOE sites, numerous contaminated units within these sites cannot be made safe for unrestricted release. Moreover, at many of these sites radiological and hazardous contaminants are already migrating, or can be expected to migrate, beyond site boundaries.

- The challenge for long-term site institutional management is therefore to assure that risks posed by such migration are successfully managed, not only within site boundaries, but in nearby areas where site managers will likely have less effective control.

2. Engineered Barriers Have Limited Lives. Engineered barriers have limited design lives compared with the time periods over which wastes will remain hazardous, and hence, will require ongoing surveillance and maintenance, and in some cases periodic replacement, to assure their continued ability to isolate wastes.

- Designing and maintaining long-lasting engineered barriers pose significant challenges for long-term institutional management. Such a system must direct attention to research and development aimed at improving the performance of both the physical systems that isolate wastes from the environment and the human institutions upon which the long-term effectiveness and monitoring of engineered barriers depends.

3. Institutional Controls Will Fail. Institutional controls and other stewardship activities are being heavily relied upon in the DOE planning for long-term management of sites where hazards will remain. Past experience with such measures suggests, however, that failures are likely to occur, possibly in the near term, and that humans and environmental resources will be put at risk as a result. The circumstances under which stewardship measures fail need to be better understood than they are at present, as must the risks associated with such failures.

- There is a need to carry out more systematic research on the types of institutional forms and incentive structures that have shown greater reliability to date, and to develop and put into place those forms and structures that appear to have the greatest promise for at least reasonably vigilant oversight and stewardship over a period of decades to a century or more. Some of the problems can be expected to last for many thousands of years. Although one might conclude that stewardship might be an alternative to cleanup or isolation for periods of tens to a few hundred years, it may not be an acceptable approach for multi-thousand-year problems.

4. Conduct "Institutional Performance Assessments." While risk assessments have been used extensively to guide cleanup decisions at DOE sites, they appear less well suited to the quantitative assessment of alternative long-term disposition than strategies that rely on stewardship measures.

- There is a need to develop the techniques and data so that what might be called "institutional systems performance assessments" can be conducted, using the same conceptual approach as for technical performance assessments; the means to do so do not exist at present. As in the scientific and technical arena, where lack of data, understanding, and analytical capability often limit the utility of risk assessment, the risk of failure of management strategies involving long-term stewardship measures is likely dominated by the contribution of "unknown unknowns"—that is, flaws in aspects of management systems that are unknown or unrecognized by analysts at the time risk assessments are conducted. It might be argued that "institutional system performance assessments" should be subjected to the same standards applied to analysis of the performance of technical systems, particularly if institutional control is viewed as an alternative to a technical remedy. For example, a performance assessment of a high-level waste repository must consider any feature, event, or process having a probability of greater than 1 in 10,000 of occurring in a 10,000-year period. If institutional performance assessments are required to meet less stringent standards, it could tend to make institutional options appear to be less "risky" than technical remediation options simply because the scope of potential risks considered is narrower in the institutional case.

5. Remediation Efforts Do Not Always Account for Long-Term Institutional Management Needs. Many remediation efforts at DOE sites, though oriented toward future or end states, in fact aim to achieve interim or temporary cleanup goals.

- Remediation planning at individual DOE waste sites is not currently occurring in a way that explicitly takes into account the needs and limitations of long-term stewardship. In particular, site planning now occurring across the DOE complex does not adequately or realistically consider the weaknesses of some stewardship measures. It is very likely that, for at least some sites where DOE is currently "completing" cleanups in the sense described at the beginning of this report, future remediation planning to revisit contamination problems will prove necessary to assure the degree of protection originally intended.

6. Present Remediation Should Aim to Facilitate Possible Re-Remediation. Actions taken today should aim to maximize chances that future generations have the *capacity* to identify and attend to unanticipated problems should they emerge at sites requiring long-term institutional management. Necessary steps include (a) assuring that the scope and severity of waste-site problems that might emerge in the future are as minimal as we can reasonably make them today, (b) assuring that future waste-site stewards have adequate resources (scientific, technical, organizational, financial, and informational) to take action when deemed necessary, and (c) assuring that monitoring and surveillance systems put in place today have (and retain) the capacity to detect unfavorable changes in site conditions at the earliest possible time.

7. Models Used in Remediation Decisions Are Inadequate. Remediation decisions at many DOE sites are relying on modeled estimates of long-term contaminant transport. The modeling approaches currently in use are often not "state of the art," nor have they been systematically reviewed to ensure the appropriateness of the assumptions used in generating these estimates. Moreover, the current "state of the art" (in environmental modeling and related computational science and technology) does not adequately capture the complicated reality that must be dealt with at many DOE sites.

8. Basic Research is Needed to Improve Long-Term Remediation Effectiveness. Whether modeling is relied upon or not, greater emphasis should be placed on developing long-term decision strategies that will be robust and adaptive in the face of actual results turning out differently than originally intended. This will not occur without concerted attention to basic research that improves understanding of the actual rather than the modeled environments at DOE sites, particularly the subsurface environments, and the dynamics of contaminant transport within them.

9. Assessment of Long-Term Impacts of Private-Sector Reindustrialization is Needed. The use of private-sector reindustrialization at DOE sites, while in many ways laudable, needs to be examined from a long-term institutional management perspective that takes into account the inherent fallibility of stewardship measures and other limitations in society's ability to manage contaminated sites over the long term.

RECOMMENDATIONS

Early and careful planning is necessary in several arenas to begin the process of developing an institutional management program for long-term disposition of radioactive and hazardous waste sites that ensures the protection of the public and the environment. The committee recommends that DOE commit the time and funding needed to develop and implement effective plans devoted to five key principles: 1) plan for uncertainty, 2) plan for fallibility, 3) develop appropriate and substantive incentive structures, 4) undertake scientific, technical, and social research and development, and 5) plan to maximize follow-through on phased, iterative, and adaptive long-term approaches.

Plan for Uncertainty

Cleanup strategies and "post-remediation" site management planning strategies must be able to adapt to a wide range of variation in possible outcomes. It is far more sensible to anticipate a wide range of possible outcomes than to focus first on defining a "most likely" outcome and then to add uncertainty by applying uncertainty ranges. Remediation strategies that avoid foreclosing options for dealing with possible outcomes often should deserve preference over those that do not.

Plan for Fallibility

Other things being equal, contaminant reduction and removal should be preferred over contaminant isolation, and either is preferable to the imposition of stewardship measures that have a high risk of failure. Stated more simply, a precautionary approach, that is, one that is self-consciously risk averse and therefore takes remedial actions even when harm is not clearly demonstrated, argues for erring on the side of contaminant reduction and removal to safer locations. Strategies that maximize the visibility of sites with residual contamination (e.g., nuclear historic parks) are preferred over those that, by making sites less apparent to the public, increase the chances that knowledge of potential risks will eventually be lost. For land containing potentially hazardous residual contamination, it is best to encourage uses that are likely to be relatively constant through time (e.g., ecological reserves) over those likely to be subject to frequent change (e.g., commercial development).

Far greater effort needs to be made to assure that information about contaminated sites is preserved and communicated effectively to future site users. It is important that relevant records on residual contaminants, remedial actions taken, and technologies used be preserved in forms that will remain accessible and readily understood by future generations. At present, it is not possible to assure that this will happen, but the committee recommends devoting greater energy to the task. Disseminate information broadly to all interested parties, and rely where possible on institutions with proven ability and positive incentives to preserve records over the long term.

Much has been published in the academic literature in recent years of the notion of "government failure." Government may work best when serving as a referee between parties of equal power (government's adjudicative function), but less well when it serves as a partisan "player." An agency's "organizational culture" is an additional consideration. Agencies that develop effective mechanisms for accomplishing one mission frequently find it very difficult to shift to other tasks. The novel demands of long-term institutional management suggest that many existing agencies, including DOE, can be expected to have difficulties adjusting to the expectations embodied in the long-term institutional management construct. This places a premium on assuring that at least some elements of future stewardship are developed from the ground up, rather than simply creating a long-term institutional management task within the structure of an existing agency.

Develop Appropriate and Substantive Incentive Structures

One of the reasons for institutional failure over time may well be that future institutions and their employees will face pressures or respond to incentives that were not adequately understood or anticipated when the institutions were first put in place. Efforts should be put forth to identify and examine the nature of incentive systems within whatever types of institutions that seem to have long-term successes at maintaining their missions, and to ask whether similar incentive structures can be built into stewardship organizations and systems.

In virtually all cases, however, stable long-term funding mechanisms and access to other needed resources appear to be necessary components of incentive structures that maintain institutional focus and effectiveness through time. Trust funds are one mechanism of this type worthy of exploration in that they could reduce susceptibility to future budget cuts from Congress or other governing bodies. Active citizen oversight of long-term management should be likewise encouraged, with stable funding or financial rewards for detection of lapses in the stewardship system, although it should never be relied upon as the lynchpin of a long-term institutional management program, since citizen groups also suffer "atrophy of vigilance."

Undertake Scientific, Technical, and Social Research and Development

Effective long-term institutional management of waste sites requires attention to limits in knowledge wherever they inhibit our ability to apply the three sets of institutional management measures discussed in the report. This requires attention to both the basic science and technology needs and the organizational and human performance aspects of long-term planning systems. Continuing efforts already underway to improve or demonstrate the short-term performance of engineered barriers are of obvious importance for extrapolating performance to a long-term institutional management system. The likelihood that such barriers will eventually fail suggests a need for greater emphasis on the specific elements of research that improve our ability to detect and correct failures of barriers once in place. The likelihood that engineered barriers will fail also highlights the need for basic research aimed at improving understanding of the surface and subsurface environments at waste sites and the dynamics of residual contaminants within them. Where transport modeling is relied upon to estimate the extent of plume migration, the errors in prediction associated with simplifying assumptions and lack of fundamental knowledge take on greater importance when very long-term planning is required. This in turn highlights the need for more attention to basic science aimed at improved understanding of the surface and subsurface environments. Attention must be given to ways to make planning more robust to accommodating "surprises" that change our understanding of residual contaminant behavior.

Plan to Maximize Follow-Through on Phased, Iterative, and Adaptive Long-Term Approaches

A long-term institutional management framework should not be static. Adaptation to changing conditions or unexpected outcomes can only be possible if the overall strategy is iterative, but iteration works best if plans have been laid for follow through on successive phases. In essence, long-term institutional management needs to be oriented toward collaborative, adaptive learning by systematically and actively seeking opportunities that cause learning and rethinking to occur.

References Cited

Alm, A.L. 1997 (December 16). Letter from Department of Energy Assistant Secretary for Environmental Management to U.S. Senator Ted Stevens submitting Closure Fund Project Management Plan, 9 pp. Department of Energy Environmental Management Office, Washington, D.C.

Applegate, J.S., and S. Dycus. 1998 (November). Institutional Controls or Emperor's Clothes? Long Term Stewardship on the Nuclear Weapons Complex. The Environmental Law Reporter ELR News & Analysis 28(11):10631-10652.

Bardach, E. 1977. The Implementation Game: What Happens After a Bill Becomes a Law. MIT Press, Cambridge, Mass.

Benford, G. 1999. Deep Time: How Humanity Communicates Across Millennia. Bard Books, New York, N.Y. 192 pp.

Bernero, R.M. 1993 (March 2). Letter from R.M. Bernero, Director, Office of Nuclear Materials Safety and Safeguards, U.S. Nuclear Regulatory Commission, to J. Lytle, Deputy Assistant Secretary for Waste Operations, Office of Waste Management, U.S. Department of Energy, Washington, D.C.

Bjornstad, D.J., D.W. Jones, and C.L. Dümmer. 1997. DOE-EM Privatization and the 2006 Plan: Principles for Procurement Policies and Risk Management: Technology. Journal of the Franklin Institute 334A:495-501.

Block, F. 1987. Revising State Theory: Essays in Politics and Postindustrialism. Temple University Press, Philadelphia.

Busenberg, G. 1999. The Evolution of Vigilance: Disasters, Sentinels and Policy Change. Environmental Politics 8(4):90-109.

Clarke, L.B. 1993. The Disqualification Heuristic: When do Organizations Misperceive Risk? Research in Social Problems and Public Policy 5:289-312.

Clarke, L.B. 1999. Mission Improbable: Using Fantasy Documents to Tame Disaster. University of Chicago Press, Chicago.

Conaway, J.G., R.J. Luxmoore, J.M. Matuszek, and R.O. Patt. 1997 (April). Tank Waste Remediation System Vadose Zone Contamination Issue: Independent Expert Panel Status Report. DOE/RL-97-49 Rev. 0, Richland, Wash.

Conner, J. R. 1990. Chemical Fixation and Solidification of Hazardous Waste. Van Nostrand Reinhold, New York, N.Y.

Dunlap, R.E., M.E. Kraft, and E.A. Rosa. 1993. Public Reactions to Nuclear Waste: Citizens' Views of Repository Siting. Duke University Press, Durham, N.C.

Easterling, D., and H. Kunreuther. 1995. The Dilemma of Siting a High-Level Nuclear Waste Repository. Kluwer, Boston, Mass.

Edelman, M. 1964. The Symbolic Uses of Politics. University of Illinois Press, Chicago.

English, M.R., D.L. Feldman, R. Inerfeld, and J. Lumley. 1997 (July). Institutional Controls at Superfund Sites: A Preliminary Assessment of Their Efficacy and Public Acceptability. Joint Institute for Energy and Environment, Knoxville, Tenn.

English, M.R., and R. Inerfeld. 1999. Institutional controls for contaminated sites: Help or hazard? Risk: Health, Safety & Environment, Spring:121-138.

Erikson, K.T. 1994. A New Species of Trouble: Explorations in Disaster, Trauma, and Community. Norton, New York, N.Y.

Erikson, K.T., E.W. Colglazier, and G.F. White. 1994. Nuclear Waste's Human Dimension. Forum for Applied Research and Public Policy (Fall):91-97.

Freudenburg, W.R. 1991. A 'Good Business Climate' as Bad Economic News? Society and Natural Resources 3:313-331.

Freudenburg, W.R. 1992. Nothing Recedes Like Success? Risk Analysis and the Organizational Amplification of Risks. Risk 3 (1-Winter):1-35.

Galanter, M. 1974. Why the 'Haves' Come Out Ahead: Speculations on the Limits of Legal Change. Law and Society Review 9:95-160.

Gee, G.W., and N.R. Wing, ed. 1994. In-Situ Remediation: Scientific Basis for Future Technologies. Thirty-Third Hanford Symposium on Health and the Environment, November 7-11, Pasco, WA. Battelle Press, Richland, Wash.

Gerber, M.S. 1992. On the Home Front: The Cold War Legacy of the Hanford Nuclear Site. University of Nebraska Press, Lincoln.

Gibbs, L.M. 1982. Love Canal: My Story. State Univ. of New York Press, Albany.

Gusterson, H. 1992. Coming of Age in a Weapons Lab: Culture, Tradition and Change in the House of the Bomb. The Sciences (May/June):16-22.

Hardert R. 1993. Public Trust and Governmental Trustworthiness: Nuclear Deception at the Fernald, Ohio, Weapons Plant. Research in Social Problems and Public Policy 5:125-148.

Harley, N.H. 2000. Back to background: Natural radiation and radioactivity exposed. Health Physics 79:121-128.

Hersh, R., K. Probst, K. Wernstedt, and J. Mazurek. 1997 (June). Linking Land Use and Superfund Cleanups: Uncharted Territory. Center for Risk Management, Resources for the Future, Washington, D.C. 107 pp.

Herzik, E.B., and A.H. Mushkatel, eds. 1993. Problems and Prospects for Nuclear Waste Disposal Policy. Greenwood Press, Westport, CT.

Hooks, G.M. 1991. Forging the Military-Industrial Complex: World War II's Battle of the Potomac. University of Illinois Press, Urbana, IL.

ICF Kaiser. 1998 (March). Managing Data for Long-Term Stewardship: Working Draft. ICF Kaiser Consulting Group, Washington, D.C.

International Atomic Energy Agency. 1976. Effects of Ionizing Radiation on Aquatic Organisms and Ecosystems. Technical Report Series No. 172, Vienna, Austria.

International Atomic Energy Agency. 1979. Methodology for Assessing the Impacts of Radioactivity on Aquatic Ecosystems. Technical Report Series No. 190, Vienna, Austria.

Jacob, G. 1990. Site Unseen: The Politics of Siting a Nuclear Waste Repository. University of Pittsburgh Press, Pittsburgh.

Jaworowski, Z. 1999 (September). Radiation risk and ethics. Physics Today 52(9):24-29.

Jones, G. 1998. Nuclear waste: management problems at the Department of Energy's Hanford spent fuel storage project. Statement dated April 12, 1998 before the Subcommittee on Oversight and Investigations, Committee on Commerce, U.S. House of Representatives. United States General Accounting Office, Washington, D.C.

Kersting, A.B., D.W. Efurd, D.L. Finnegan, D.J. Rokop, D.K. Smith, and J.L. Thompson.1999 (January 7). Migration of plutonium in ground water at the Nevada Test Site. Nature 397:56-59.

Korngold, Gerald. 1984. Privately Held Conservation Servitudes: A Policy Analysis in the Context of Gross Real Covenants and Easements. Texas Law Review 63, 433-495.

Krannich, R.S., and A.E. Luloff. 1991. Problems of Resource Dependency in U.S. Rural Communities. Progress in Rural Policy and Planning 1:5-18.

LaPorte, T.R. 1996. Highly Reliable Organizations: Unlikely, Demanding, and At Risk. Journal of Crisis and Contingency Management 4(2):60-71.

LaPorte, T.R., and A. Keller. 1996. Assuring Institutional Constancy: Requisite for Managing Long-Lived Hazards. Public Administration Review 56(6):535-544.

LaPorte, T.R., and D. Metlay. 1996. Facing Deficit of Trust: Hazards and Institutional Trustworthiness. Public Administration Review 54(4):342-347.

Lawless, W.F. 1991. A Social Psychological Analysis of the Practice of Science: The Problem of Military Nuclear Waste Management, p. 91-96. In R.G. Post [ed.] Waste Management '91: Waste Processing, Transportation, Storage and Disposal, Technical Programs and Public Education (Vol. 2). University of Arizona, Tucson.

Levine, A. 1982. Love Canal: Science, Politics, and People. Lexington Books: Lexington, Mass.

Lodwick, D.G. 1993. Rocky Flats and the Evolution of Distrust. Research in Social Problems and Public Policy 5:149-170.

Logan, J.R., and H.L. Molotch. 1987. Urban Fortunes: The Political Economy of Place. University of California Press, Berkeley.

Martin, B. 1999. Suppression of dissent in science. Research in Social Problems and Public Policy 7:105-135.

Mazur, A. 1998. A Hazardous Inquiry: The Rashomon Effect at Love Canal. Harvard University Press, Cambridge, Mass.

McCoy, C. 1989. Broken Promises: Alyeska Record Shows How Big Oil Neglected Alaskan Environment. Wall Street Journal (July 6):A1, A4.

Mojtabai, A.G. 1986. Blessed Assurance: At Home with the Bomb in Amarillo, Texas. Houghton Mifflin, Boston, Mass.

Molotch, H. 1976. Oil in Santa Barbara and Power in America. Sociological Inquiry 40(Winter):131-144.

Morgan, G. 1986. Images of Organization. Sage, London.

Morone, J.G., and E.J. Woodhouse. 1989. The Demise of Nuclear Energy? Lessons for Democratic Control of Technology. Yale University Press, New Haven, Conn.

Mulvernon, N. 1998 (April 12). The Sign of Stewardship: A Case Study in Stewardship. Oak Ridge Reservation End Use Working Group, Oak Ridge, Tenn.

National Council on Radiation Protection and Measurement 1991. Effects of Ionizing Radiation on Aquatic Organisms. NCRP Report No. 109, Bethesda, Md.

National Council on Radiation Protection and Measurement. 1995. Principles and Application of Collective Dose in Radiation Protection. NCRP Report No. 121, Bethesda, Md.

National Council on Radiation Protection and Measurements. 1999 (30 June). Recommended Screening Levels for Contaminated Surface Soil and Review of Factors Relevant to Site Specific Studies. National Council on Radiation Protection and Measurements (NCRP) Report 129, Bethesda, Md.

REFERENCES CITED

National Research Council. 1983. Risk Assessment in the Federal Government: Managing the Process. Committee on the Institutional Means for Assessment of Risks to Public Health, National Academy Press, Washington, D.C. 191 pp.

National Research Council. 1989. Improving Risk Communication. Committee on Risk Perception and Communication, National Academy Press, Washington, D.C. 332 pp.

National Research Council. 1994a. Science and Judgement in Risk Assessment. Committee on Risk Assessment of Hazardous Air Pollutants, National Academy Press, Washington, D.C.

National Research Council. 1994b. Building Consensus Through Risk Assessment and Management of the Department of Energy's Environmental Remediation Program. National Academy Press, Washington, D.C.

National Research Council. 1994c. Alternatives for Ground Water Cleanup. Committee on Ground Water Cleanup Alternatives, National Academy Press, Washington, D.C. 315 pp.

National Research Council. 1995a. Safety of the High-Level Uranium Ore Residues at the Niagara Falls Storage Site, Lewiston, New York. Committee on Remediation of Buried and Tank Wastes, Washington, D.C. 73 pp.

National Research Council. 1995b. Improving the Environment: An Evaluation of DOE's Environmental Management Program. Committee to Evaluate the Science, Engineering, and Health Basis of the Department of Energy's Environmental Management Program, National Academy Press, Washington, D.C. 211 pp.

National Research Council. 1995c. Technical Bases for Yucca Mountain Standards. Committee on Technical Bases for Yucca Mountain Standards, National Academy Press, Washington, D.C. 205 pp.

National Research Council. 1996a. Understanding Risk: Informing Decisions in a Democratic Society. Committee on Risk Characterization, National Academy Press, Washington, D.C. 249 pp.

National Research Council. 1996b. Nuclear Wastes: Technologies for Separations and Transmutation. Committee on Separations Technology and Transmutation Systems, National Academy Press, Washington, D.C. 571 pp.

National Research Council. 1996c. Barriers to Science: Technical Management of the Department of Energy Environmental Remediation Program. Committee on Remediation of Buried and Tank Wastes, National Academy Press, Washington, D.C. 23 pp.

National Research Council. 1996d. The Hanford Tanks: Environmental Impacts and Policy Choices. Committee on Remediation of Buried and Tank Wastes, National Academy Press, Washington, D.C. 73 pp.

National Research Council. 1996e. Environmental Management Technology-Development Program at the Department of Energy: 1995 Review. Committee on Environmental Management Technologies, National Academy Press, Washington, D.C. 131 pp.

National Research Council. 1997a. Barrier Technologies for Environmental Management: Summary of a Workshop. Committee on Remediation of Buried and Tank Wastes, National Academy Press, Washington, D.C.

National Research Council. 1997b. Innovations in Ground Water and Soil Cleanup: From Concept to Commercialization. Committee on Innovative Remediation Technologies, National Academy Press, Washington, D.C. 292 pp.

National Research Council. 1997c. Building an Effective Environmental Management Science Program: Final Assessment. Committee on Building an Environmental Management Science Program, National Academy Press, Washington, D.C. 66 pp.

National Research Council. 1998a. Systems Analysis and Systems Engineering in Environmental Remediation Programs at the Department of Energy Hanford Site. Committee on Remediation of Buried and Tank Wastes, National Academy Press, Washington, D.C. 51 pp.

National Research Council. 1998b. A Review of Decontamination and Decommissioning Technology Development Programs at the Department of Energy. Committee on Decontamination and Decommissioning, National Academy Press, Washington, D.C. 88 pp.

National Research Council. 1998c. Peer Review in Environmental Technology Development Programs: The Department of Energy's Office of Science and Technology. Committee on the Department of Energy—Office of Science and Technology's Peer Review Program, National Academy Press, Washington, D.C. 114 pp.

National Research Council. 1998d. Study of the Decision Processes Related to Long-Term Disposition of U.S. Department of Energy Waste Sites and Facilities: Interim Status Report. Committee on Remediation of Buried and Tank Wastes, National Academy Press, Washington, D.C. 13 pp.

National Research Council. 1999a. Environmental Clean-Up at Navy Facilities: Risk-Based Methods. Committee on Environmental Remediation at Naval Facilities, National Academy Press, Washington, D.C. 143 pp.

National Research Council. 1999b. Decision Making in the U.S. Department of Energy's Environmental Management Office of Science and Technology. Committee on Prioritization and Decision Making in the Department of Energy Office of Science and Technology, National Academy Press, Washington, D.C. 215 pp.

National Research Council. 1999c. Technologies for Environmental Management: The Department of Energy's Office of Science and Technology. Board on Radioactive Waste Management, National Academy Press, Washington, D.C. 73 pp.

National Research Council. 1999d. An End State Methodology for Identifying Technology Need for Environmental Management, with an Example from the Hanford Site Tanks. Committee on Technologies for Cleanup of High-Level Waste in Tanks in the DOE Weapons Complex. National Academy Press, Washington, D.C. 93 pp.

National Research Council. 1999e. Groundwater & Soil Cleanup: Improving Management of Persistent Contaminants. Committee on Technologies for Cleanup of Subsurface Contaminants in the DOE Weapons Complex. National Academy Press, Washington, D.C. 285 pp.

National Research Council. 2000a. Natural Attenuation for Groundwater Remediation. Committee on Intrinsic Remediation, National Academy Press, Washington, D.C.

National Research Council. 2000b. Research Needs in Subsurface Science: U.S. Department of Energy's Environmental Management Science Program. Committee on Subsurface Contamination at DOE Complex Sites, National Academy Press, Washington, D.C. 159 pp.

Ohm, B.W. 2000. The purchase of scenic easements and Wisconsin's Great River Road: A progress report on perpetuity. Journal of the American Planning Association 66(2):177-188.

Peach, J.D. 1988. Ineffective Management and Oversight of DOE's P-Reactor at Savannah River, S.C., Raises Safety Concern" Statement before the Committee on Governmental Affairs, United States Senate, Subcommittee on Environment, Energy, and Natural Resources, Committee on Government Operations, House of Representatives. United States General Accounting Office, Washington, D.C.

Pendergrass, J. 1996 (March). Use of Institutional Controls as Part of a Superfund Remedy: Lessons from Other Programs. Environmental Law Reporter 26:10109-10123.

Perrow, C. 1984. Normal Accidents: Living with High-Risk Technologies. Basic Books, New York, N.Y.

President's Commission on the Accident at Three Mile Island. 1979. The Need for Change: The Legacy of Three Mile Island. U.S. Government Printing Office, Washington, D.C.

Probst, K.N., and A.I. Lowe. 2000 (January). Cleaning Up the Nuclear Weapons Complex: Does Anybody Care? Center for Risk Management, Resources for the Future, Washington D.C. 38 pp.

Probst, K.N., and M.H. McGovern. 1998 (June). Long-Term Stewardship and the Nuclear Weapons Complex: The Challenge Ahead. Center for Risk Management, Resources for the Future, Washington D.C. 67 pp.

Richland Environmental Restoration Project and Bechtel Hanford, Inc. 1999 (February). Submittal for 1998 Project of the Year—C Reactor Interim Safe Storage. Richland, Wash.

Robison, W.L., K.T. Bogen, and C.L. Conrado. 1997. An Updated Dose Assessment for Resettlement Options at Bikini Atoll: a U.S. Nuclear Test Site. Health Physics 73:100-114.

Rochlin, G.I. 1996. Reliable Organizations: Present Research and Future Directions. Journal of Crisis and Contingency Management 4(2):55-59.

Rosa, E.A., and D.L. Clark, Jr. 1999. Historical Routes to Technological Gridlock: Nuclear Technology as Prototypical Vehicle. Research in Social Problems and Public Policy 7:21-57.

Rumer, R.R., and J.K. Mitchell, eds. 1996. Assessment of Barrier Containment Technologies: A Comprehensive Treatment for Environmental Remediation Applications. National Technical Information Service, Springfield, Va.

Rumer, R.R., and M.E. Ryan, eds. 1995. Barrier Containment Technology for Environmental Remediation Applications. J. Wiley & Sons, Inc., New York, N.Y.

Russell, M. 1997. Toward a Productive Divorce: Separating DOE Cleanups from Transitional Assistance. The Joint Institute for Energy and Environment, Knoxville, Tenn.

Rust Geotech. 1996. Vadose Zone Monitoring Project at the Hanford Tank Farms: Tank Summary Data Report for Tank SX-112. Grand Junction Projects Office Report GJ-HAN-14, Tank SX112, Grand Junction, Colo.

Sagan, S. 1994. Toward a Political Theory of Reliability. Journal of Crisis and Contingency Management 2(4):228-240.

Secretary of Energy Advisory Board Task Force on Radioactive Waste Management. 1993. Earning Public Trust and Confidence: Requisites for Managing Radioactive Wastes. Final Report of the Secretary of Energy Advisory Board Task Force on Radioactive Waste Management, Washington, D.C.

Sheak, R.J., and P. Cianciolo. 1993. Notes on Nuclear Weapons Plants and their Neighbors: The Case of Fernald. Research in Social Problems and Public Policy 5:97-122.

Short, J.F., and L. Clarke, eds. 1992. Organizations, Uncertainties, and Risk. Westview, Boulder.

Shrader-Frechette, K.S. 1993. Risk Methodology and Institutional Bias. Research in Social Problems and Public Policy 5:207-223.

Slovic, P. 1991. Perceived Risk, Trust, and the Politics of Nuclear Waste. Science 254:1603-1607.

Slovic, P. 1993. Perceived Risk, Trust, and Democracy. Risk Analysis 13 (6-February):675-682.

Stone, C.N. 1989. Regime Politics: Governing Atlanta, 1946-1988. University Press of Kansas, Lawrence.

Tangley, L. 1998. U.S. News & World Report, February 16.

United Nations Scientific Committee on the Effects of Atomic Radiation. 1993. UNSCEAR 1994 Report to the General Assembly, with Scientific Annexes. United Nations, New York, N.Y.

United Nations Scientific Committee on the Effects of Atomic Radiation. 1996. Annex on Effects of Radiation on the Environment. United Nations, New York, N.Y.

U.S. Atomic Energy Commission. 1974 (June). Regulatory Guide 1.86, Termination of Operating Licenses for Nuclear Reactors. Washington, D.C.

U.S. Congress Office of Technology Assessment. 1991 (February). Complex Cleanup: The Environmental Legacy of Nuclear Weapons Production. Office of Technology Assessment OTA-0484, U.S. Government Printing Office, Washington, D.C. 212 pp.

U.S. Department of Defense, U.S. Department of Energy, U.S. Environmental Protection Agency, and U.S. Nuclear Regulatory Commission. 1997 (December). Multi-Agency Radiation Survey and Site Investigation Manual (MARSSIM): Final. NUREG-1575, EPA 402-R-97-016, Washington, D.C.

U.S. Department of Energy. 1995a (March). Estimating the Cold War Mortgage: The 1995 Baseline Environmental Management Report. Office of Environmental Management, DOE/EM-0232, Washington, D.C.

U.S. Department of Energy. 1995b (November 17). DOE Memorandum: Application of DOE 5400.5 Requirements for Release and Control of Property Containing Residual Radioactive Material, EH-412. Air, Water, and Radiation Division, Washington, D.C.

U.S. Department of Energy. 1996 (June). The 1996 Baseline Environmental Management Report. Office of Environmental Management, DOE/EM-0290, Washington, D.C.

REFERENCES CITED

U.S. Department of Energy. 1997a (February 19). Long-Term Stewardship of DOE Sites: Scope of Work and Considerations for Success. Office of Strategic Planning and Analysis Draft Report, Washington, D.C.

U.S. Department of Energy. 1997b. Linking Legacies: Connecting the Cold War Nuclear Weapons Processes to Their Environmental Consequences. Office of Environmental Management DOE/EM-0319, Washington, D.C. 230 pp.

U.S. Department of Energy. 1998a (June). Accelerating Cleanup: Paths to Closure. Office of Environmental Management DOE/EM-0362, Washington, D.C.

U.S. Department of Energy. 1998b (September). Response II.A to U.S. Nuclear Regulatory Commission Comments on SRS HLW Tank Closure. Savannah River Operations Office, Aiken, S.C.

U.S. Department of Energy. 1998c (December). Groundwater/Vadose Zone Integration Project Specification. Richland Operations Office DOE/RL-98-48, Richland, Wash.

U.S. Department of Energy. 1999 (October). From Cleanup to Stewardship: A Companion Report to *Accelerating Cleanup: Paths to Closure* and Background Information to Support the Scoping Process Required for the 1998 PEIS Settlement Study. Office of Environmental Management DOE/EM-0466, Washington, D.C.

U.S. Department of Energy. 2000a (January 24). Guidance for the Development of the FY 2000 National Defense Authorization Act (NDAA) Long-Term Stewardship Report. Office of Environmental Management, Washington, D.C. 28 pp.

U.S. Department of Energy. 2000b (March). Status Report on Paths to Closure. Office of Environmental Management DOE/EM-0526, Washington, D.C.

U.S. Environmental Protection Agency. 1998a (March). Institutional Controls: A Reference Manual. U. S. Environmental Protection Agency Draft Report. Washington, D.C.

U.S. Environmental Protection Agency. 1998b (April 21). Memorandum, EPA Region IV Policy Assuring Land Use Controls at Federal Facilities. Washington, D.C.

U.S. Environmental Protection Agency. 1998c. Memorandum, EPA Region 10 Final Policy on the Use of Institutional Controls at Federal Facilities. Washington, D.C.

U.S. Environmental Protection Agency. 1999 (September 30). Follow-Up Audit of Superfund Five-Year Review Program. Office of Inspector General Audit Report 1999-P-218. Washington, D.C.

U.S. Environmental Protection Agency. 2000. Institutional Controls and Transfer of Real Property under CERCLA Section 120(h)(3)(A), (B), or (C). Washington, D.C.

U.S. General Accounting Office. 1989. Nuclear Health and Safety: Savannah River's Unusual Occurrence Reporting Program Has Been Ineffective. U.S. General Accounting Office, Washington, D.C.

U.S. General Accounting Office. 1993. Nuclear Waste: Hanford Tank Waste Program Needs Cost, Schedule, and Management Changes. Report to the Chairman, Environment, Energy, and Natural Resources Subcommittee, Committee on Government Operations, House of Representatives. U.S. General Accounting Office, Washington, D.C.

U.S. General Accounting Office. 1996. Nuclear Waste: Management and Technical Problems Continue to Delay Characterizing Hanford's Tank Waste. Report to the Secretary of Energy. U.S. General Accounting Office, Washington, D.C.

U.S. General Accounting Office. 1997a (February). Department of Energy Contract Management. GAO High-Risk Series GAO/HR-97-13, Washington, D.C.

U.S. General Accounting Office. 1997b (July). Department of Energy's Project to Clean Up Pit 9 at Idaho Falls is Experiencing Problems. GAO/-RCED-97-180, Washington, D.C. 35 pp.

U.S. General Accounting Office. 1998 (May). Department of Energy: Alternative Financing and Contracting Strategies for Cleanup Projects. GAO/RCED-98-169, Washington, D.C.

U.S. General Accounting Office. 1999 (April). Nuclear Waste: DOE's Accelerated Cleanup Strategy Has Benefits but Faces Uncertainties. GAO/RCED-99-129, Washington, D.C. 21 pp.

U.S. Nuclear Regulatory Commission. 1998 (August). USNRC Draft Regulatory Guide, DG 4006, Demonstrating Compliance with the Radiological Criteria for License Termination. Washington, D.C.

U.S. Nuclear Regulatory Commission. 1999 (December 15). Classification of Savannah River Residual Waste as Incidental. SECY-99-284, Washington, D.C.

U.S. Nuclear Regulatory Commission. (in review). Radiological Assessments for Clearance of Equipment and Materials from Nuclear Facilities. NUREG-1640, Vol. 1, pp. 2-8. Washington, D.C.

U.S. Secretary of Energy Advisory Board Task Force on Radioactive Waste Management. 1993. Earning Public Trust and Confidence: Requisites for Managing Radioactive Wastes. U.S. Department of Energy, Washington, D.C.

Vaughan, D. 1997. The Challenger Launch Decision: Risky Technology, Culture, and Deviance at NASA. University of Chicago Press, Chicago.

Wald, M. 1992. Rules rewritten on nuclear waste. New York Times, October 11, p. 13.

The Washington Advisory Group. 1999 (December 17). Managing Subsurface Contamination: Improving Management of the Department of Energy's Science and Engineering Research on Subsurface Contamination: Final Report. Washington, D.C.

Weimer, D.L., and A.R. Vining. 1999. Policy Analysis: Concepts and Practice. Third Edition. Prentice-Hall, Upper Saddle River, N.J. 486 pp.

Weinberg, A. 1972 (April 2). Science and Trans-Science. Minerva 10:209-222.

Weisgall, J.M. 1994. Operations Crossroad: The Atomic Tests at Bikini Atoll. Naval Institute Press, Annapolis, MD. 440pp.

Wilson D.J., and A.N. Clarke. 1994. Hazardous Waste Site Soil Remediation – Theory and Applications of Innovative Technology. Marcel Dekker Inc., New York, N.Y.

APPENDIXES

APPENDIX A

Committee's Statement of Task

The purpose of this project is to assess approaches for developing criteria for transition from active to passive remediation and subsequent long-term disposition, including institutional control with monitoring and surveillance, of U.S. Department of Energy waste sites and facilities such as Hanford, Washington; Savannah River, South Carolina; Idaho National Engineering Laboratory; and Oak Ridge National Laboratory, Tennessee. Such criteria will include technical feasibility, future land use, performance assessment of remediation activities, and risks to health, safety, and the environment associated with long-term site disposition. Relevant federal and state regulatory requirements and agreements will be included. Appropriate approaches will be applicable to facilities such as high-level radioactive waste tanks (including related facilities and contaminated environments), buried radioactive waste (such as the Hanford low-level waste disposal sites), and on environments contaminated by nuclear testing (such as the Nevada Test Site weapons test event location).

April 1997

APPENDIX B

Closure Plans for Major DOE Sites

Raymond G. Wymer

Information for the table in this appendix was taken primarily from the following sources: *Baseline Environmental Management Reports* (U.S. Department of Energy, 1995, 1996), *Accelerating Cleanup: Paths to Closure* (U.S. Department of Energy, 1998), and *From Cleanup to Stewardship* (U.S. Department of Energy, 1999). The Department of Energy uses the term "sites" in several ways, for example, to refer to national laboratories or to installations such as the Hanford Site or the Savannah River Site. In other instances it refers to specific areas within the major sites as sites. As a consequence the number of contaminated "sites" can vary from several dozen to many hundreds, depending upon the definition used. In order to bound the number of "sites" to be considered, the "sites" listed in the accompanying table are those discussed in the above DOE reports.

REFERENCES CITED

U.S. Department of Energy. 1995 (March). Estimating the Cold War Mortgage: The 1995 Baseline Environmental Management Report. Office of Environmental Management, DOE/EM-0232, Washington, D.C.

U.S. Department of Energy. 1996 (June). The 1996 Baseline Environmental Management Report. Office of Environmental Management, DOE/EM-0290, Washington, D.C.

U.S. Department of Energy. 1998 (June). Accelerating Cleanup: Paths to Closure. Office of Environmental Management DOE/EM-0362, Washington, D.C.

U.S. Department of Energy. 1999 (October). From Cleanup to Stewardship: A Companion Report to *Accelerating Cleanup: Paths to Closure* and Background Information to Support the Scoping Process Required for the 1998 PEIS Settlement Study. Office of Environmental Management DOE/EM-0466, Washington, D.C.

TABLE B-1 Closure Plans for Major DOE Sites (Sources: U.S. Department of Energy [1995, 1996, 1998a, 1999])

State	Site	Responsibility for Site/End Use(s)	End State	Conditions of Closure	Completion Date
Alaska	Amchitka Island	Release to U.S. Fish and Wildlife Service or U.S. Bureau of Wildlife Mgt.	Greenfield on surface/ institutional control on all sub-surface areas near shot cavities	Sub-surface soil and groundwater surveillance and monitoring planned for 100 years, but assumed to be in perpetuity; will require controlled access; surface released for uncontrolled use (open space)	2001
California	Energy Technology Engineering Center (ETEC)	Site will be turned over to Boeing/Rocketdyne	Probably industrial use under surveillance and monitoring and deed restrictions	Remediation of groundwater, soils and decontamination and decommissioning (D&D) of several bldg.; residual inorganic, PCB, semivolatile organic chemical (SVOC), mercury and dioxin left in soil; contaminated soil over 1×10^{-5} disposed off site; facilities require D&D of radionuclides and sodium under RCRA	2006
California	General Atomics Site (GA)	U.S. Department of Energy (DOE) keeps liability until all waste is off the site then GA assumes site	Greenfield	GA responsible for post-remediation monitoring	1999
California	General Electric (GE) Vallecitos Nuclear Center	GE owns the site; after cleaning up hot cell and glove box DOE has no further responsibility	Brownfield/part of site will be zoned industrial; hot cell not to be used commercially	DOE will clean up hot cell #4 and glove box; Univ CA-Davis has primary responsibility for post-closure monitoring	2005
California	Laboratory for Energy-Related Health Research (LEHR)	Univ CA-Davis owns the site and is responsible for radioactive waste burial trench and 3 landfills; DOE has leased since 1958	Controlled access; decontaminated to industrial use levels	DOE responsible for decontamination of septic tanks, burial trenches, dry wells, dog pen facilities, etc.; limited institutional controls and monitoring may be necessary	2002
California	Lawrence Berkeley National Laboratory (LBNL)	Land leased to DOE by Univ CA; 134 acres adjacent to Univ CA-Berkeley:	Ongoing DOE mission	Groundwater treatment system in place by 2003; no cleanup level defined for tritium in groundwater; long-term monitoring through 2032; underground tanks will be removed	2003

111

(continued)

112

TABLE B-1 Continued

State	Site	Responsibility for Site/End Use(s)	End State	Conditions of Closure	Completion Date
California	Lawrence Livermore National Laboratory (LLNL)-Main Site	DOE will continue to own and manage site; Univ CA operates site	Brownfield/future land use to be research and industrial	Soil and groundwater remediation in progress; no solid waste disposal on site; on U.S. Environmental Protection Agency (EPA) national priority list; on-site groundwater must be remediated to EPA maximum contaminant levels (MCL); groundwater stewardship may last to 2015	2006
California	Lawrence Livermore National Laboratory (LLNL)-Site 300	LLNL will occupy indefinitely	DOE ongoing mission/ controlled access/mix of industrial and wildlife management areas	Groundwater treatment operational by 2006. Will continue until negotiated goals are met; groundwater monitoring for at least 23 years; landfills will require at least 23 years of surveillance and monitoring and cap inspections and repairs	2006
California	Sandia National Laboratories-CA	Ongoing mission under Defense Programs	Brownfield/ongoing mission	Designated solid waste mgt. areas remediated or under management controls such that no further action is necessary; remediation and waste disposal of 23 release sites by 2001	2001
California	Stanford Linear Accelerator Center (SLAC)	To be returned to DOE Office of Energy Research by end of 2000	Ongoing DOE mission	Soil and groundwater will be cleaned up; network of monitoring wells installed; soil contaminated by solvents at 10 to 20 feet will stay that way	2000
Colorado	Grand Junction Office Site	No radiological restrictions	Greenfield/industrial/ recreational	Administrative control of groundwater until verification that passive remediation has achieved cleanup goals (by approximately 2076)	2002
Colorado	Rio Blanco	DOE will maintain institutional control of sub-surface areas near shot cavities in perpetuity	Surface area will be released for alternative uses/no radiological restrictions	Site remains under controlled access; monitoring planned for 100 years, but assumed to be in perpetuity	2005
Colorado	Rocky Flats Buffer Zone	DOE may transfer site to another entity as cleanup becomes more complete	Final record of decision (ROD) will determine stewardship requirements; likely available for re-use as open space		2010 (2006)

State	Site	Status/Responsibility	End State	Actions	Year
Colorado	Rocky Flats Industrial Area (consists of six operable units; 95 individual contaminated sites)[1]		Various end states, depending upon particulars of sites	Clean up in compliance with environmental laws and regulations; surveillance and monitoring after closure of each operable unit; monitoring for greater than 30 years after site closure; subsurface facilities will be capped and left in place	1998
Colorado	Rulison	DOE will maintain institutional control of shot cavities	Surface area will be released for alternate uses	Site surface released for recreation; subsurface areas near no radiological restrictions; subsurface and groundwater remains under institutional controls; long-term surveillance and monitoring	2000
Idaho	Argonne National Laboratory-West	DOE Nuclear Energy is responsible for waste mgt. program	Ongoing DOE mission/ brownfield/industrial, commercial/residual contamination in soil and groundwater	Groundwater remediation will be ongoing for 5 years with monitoring for at least 20 years; surveillance and maintenance for about 100 years; DOE will conduct 5-year CERCLA reviews and sampling for 20 years; after DOE departs deed restrictions will be needed to maintain industrial use levels	2000
Idaho	Idaho National Engineering and Environmental Laboratory (INEEL)[2]	Currently DOE; long-range plan is to be a national multi-program engineering and environmental laboratory	Cleanup per Federal Facility Agreement and Consent Order (FFACO) and all existing and future agreements; grazing and industrial use	Industrial and open space; on-site disposal cell for contact handled (CH) low-level waste (LLW); store spent fuel until 2035; treat and store high-level waste (HLW) until 2070; no residential use for 100 years; various waste area groups (WAG) handled according to need and anticipated use	2050
Illinois	Argonne National Laboratory-East	DOE Energy Research is landlord; DOE has program responsibility for environmental restoration, stewardship, and land use	Ongoing DOE mission	On-site containment of some residual contamination; annual sampling and monitoring of soil; groundwater remedial options include pump and treat after 2002; composite caps over several landfills	2002
Iowa	Ames Laboratory	Waste mgt. program transferred to DOE Energy Research in 2000	Greenfield	All waste treated and/or disposed of off site	1999
Kentucky	Maxey Flats Disposal Site	Commonwealth of KY has long-term stewardship; permanent LLW disposal site; controlled access	Controlled access	Cleanup levels in accordance with CERCLA ROD; wastes stay on site	2002

(continued)

TABLE B-1 Continued

State	Site	Responsibility for Site/End Use(s)	End State	Conditions of Closure	Completion Date
Kentucky	Paducah Gaseous Diffusion Plant	On-going mission by US Enrichment Corp. (USEC)	Brownfield/controlled access; other land restricted industrial, open space/recreational	Contaminated burial grounds and landfills closed in place in industrial area; deed restrictions or use limitation on areas with contamination; pump and treat off-site plumes until 2070; federal government maintains stewardship forever; caps over burial grounds; soil monitoring for hundreds of years; groundwater pump and treat for at least 100 years	2010
Mississippi	Salmon Site	Site will be transferred to Mississippi	No radiological restrictions/ characterization and remediation under RCRA/use as a wilderness area	Site remains under controlled access; DOE responsible for institutional controls forever; monitoring for 100 years	1999
Missouri	Kansas City Plant	Defense Programs is landlord/commercial use	Brownfield	Dense nonaqueous-phase liquids (DNAPL) cleaned up with innovative technology; groundwater treatment and monitoring from two to hundreds of years	1999
Missouri	Weldon Springs Site	155 acres of plant site released to unrestricted use; 9-acre quarry for recreational use; 62-acre disposal cell controlled access	Greenfield and brownfield/ controlled access/engineered disposal facility with clay liner and stone cap for debris, sludge, contaminated soil, asbestos, low-level PCB	Federal government stewardship forever	2002
Multiple States	Uranium Mill Tailings Remedial Action (UMTRA) Sites	Site responsibilities vary: state, local, and DOE; restrictions range from uncontrolled access to restricted access	Greenfield/restricted access	Many sites rely on natural attenuation to reach EPA groundwater standards	Various
Nevada	Central Nevada Test Site	DOE responsible for institutional controls of sub-surface soil and contaminated groundwater	Future surface use with "no radiological restrictions"; No technology available for bomb crater cleanup; economic redevelopment possible in parts	Site remains under controlled access; in-situ containment; treatment for commercial disposal; indefinite term of monitoring	2006 (est)

State	Site	Plans	End Use	Date
Nevada	Nevada Test Site (NTS) (including Tonopah test range)	DOE ongoing mission of nuclear stockpile stewardship; federal government will own the land forever	Brownfield development possible in southwestern portion/controlled access all over/final cleanup levels to be negotiated with state regulators/waste management sites used for LLW and mixed wastes	Surface and soil plumes outside of NTS will be remediated; sub-surface contaminants in and around shots will not be remediated; filled pits and trenches will be closed and capped; modeling and monitoring in perpetuity to predict movement of radionuclides in groundwater	2014 (est.)
Nevada	Project Shoal Site	DOE will not maintain an active presence but will maintain institutional controls forever for subsurface soil and groundwater	Surface soil re-use with no radiological restrictions	Site remains under controlled access; restricted access to groundwater forever; monitoring forever	2004
New Jersey	Princeton Plasma Physics Laboratory	DOE Office of Science gets waste mgt. program in 2000	Ongoing research mission	Contaminated soil and sediment treated and disposed off site; no groundwater remediation required	1999
New Mexico	Gasbuggy	DOE will maintain institutional control of sub-surface areas near shot cavities	Surface area will be released for alternate uses with no radiological restrictions	Site remains under controlled access; monitoring forever; groundwater monitoring wells refurbished or replaced every 25 years	2005
New Mexico	Gnome-Coach	DOE will not maintain an active presence, but will maintain institutional control of sub-surface areas near shot cavities; land released without restrictions or given to Bureau of Land Mgt.	Surface area will be released for alternate uses with no radiological restrictions	Site remains under controlled access; annual monitoring planned for 100 years, but assumed to be forever	2004
New Mexico	Los Alamos National Laboratory (LANL)	Ongoing research mission; transfer up to 4650 acres to county and San Idelfonso Pueblo	Brownfield/industrial/commercial use/DOE mission	Legacy mixed LLW off site by 2004; environmental restoration project complete 2008; residual radioactive, metal and organic contamination; indefinite radiological contamination surveillance and monitoring; no groundwater remediation of regional aquifer deemed necessary; groundwater monitoring for over 30 years; contaminated material disposal areas closed with engineered barriers and long-term surveillance and monitoring	2008

(continued)

115

TABLE B-1 Continued

State	Site	Responsibility for Site/End Use(s)	End State	Conditions of Closure	Completion Date
New Mexico	Lovelace Respiratory Research Institute (LRRI)	U.S. Air Force leases the land to DOE with renewal option; Office of Science is operational landlord	Brownfield	Groundwater contamination exceeds NMED 10 Tg/L level; natural attenuation of nitrates and diesel products expected to achieve standard; surveillance and monitoring until 2006	2000
New Mexico	Sandia National Laboratories-NM	Ongoing mission under Defense Programs; industrial (DOE programmatic) uses beginning in 2001	Brownfield/ongoing mission/chemical waste and mixed waste landfills and a disposal cell will remain	Controlled access of landfills and disposal cell unless wastes are sent off site; under federal control in perpetuity; long-term institutional controls; 30 years of monitoring per RCRA	2001
New Mexico	Waste Isolation Pilot Plant (WIPP)	WIPP is neither a cleanup nor a disposal site; DOE control for ongoing waste mgt. for CH and remote handled (RH) transuranic (TRU) wastes until 2033	After completion of TRU disposal project surface area will be unrestricted for recreational and agricultural use	Monuments and markers to warn people of presence of radioactive wastes; no access to underground; 124 acres passive institutional control	2038
New York	Brookhaven National Laboratory	Office of Science is landlord; ongoing research mission	Final ROD not complete and remediation strategies are not finalized as of June 1999	Groundwater remediation and monitoring until 2031; former and current landfills capped; newly generated wastes disposed off site; institutional controls for 100 years with deed and use restrictions at site closure time	2006
New York	Separations Process Research Unit (SPRU)-Knolls	Owned by Knolls Atomic Power Laboratory	Greenfield/to be released by owner for unrestricted use	In standby since 1953; surveillance and monitoring in place; some transuranic wastes	2014
New York	West Valley Demonstration Project	Site and facilities owned by New York state and licensed by NRC; DOE manages oversight responsibilities; final end state not determined	Remediation strategy and final EIS not complete; after completion of project facility operational responsibilities will be transferred to New York Energy Research Development Authority	Unknown pending completion of final Environmental Impact Statement (EIS)	Unknown

Ohio	Ashtabula Environmental Management Project	Owned by RMI, a private company	RMI has sole responsibility for site after 2003	Future use assumed to be industrial consistent with surrounding property use and zoning; surficial soil to be remediated to <30 pCi/g; long-term sampling and monitoring of groundwater	2003
Ohio	Columbus Environmental Management Project-West Jefferson	Return to Battelle for unrestricted use by 2005	Brownfield/industrial use	Clean up for use without radiological restrictions; all waste streams to be shipped off site	2005
Ohio	Fernald Environmental Management Project (FEMP)	DOE or a successor federal agency maintains stewardship; use may be recreational or industrial	Brownfield/no residential or agricultural use; access to 138-acre on-site disposal facility restricted forever	Large on-site LLW disposal facility; controlled access; restore aquifer to 20 ppb uranium contamination; 23 acres set aside for future economic development; groundwater monitoring forever	2008 (2005)
Ohio	Miamisburg Environmental Management Project-(MEMP) (Mound)	Transfer to city of Miamisburg by 2004 except for Office of Nuclear Facilities for ongoing NE mission	Brownfield/cleanup to EPA industrial use standards	DOE retains responsibility for contaminated areas; DOE will remediate on-site groundwater to industrial use levels and off-site groundwater to residential levels; DOE has duty to conduct annual assessments of compliance with deed restrictions and to enforce compliance	2004
Ohio	Portsmouth Gaseous Diffusion Plant	USEC will use the plant for the foreseeable future; federal government responsible for stewardship forever	Brownfield/combination of mixed industrial and recreational use	Contaminated burial grounds and landfills closed in place in industrial area; deed restrictions or use limitation on areas with contamination; complete remediation of waste sites in 2035; shut down groundwater treatment in 2050 and monitoring of passive treatment in 2055; seven capped landfills remain on site	2005
South Carolina	Savannah River Site (SRS)[3]	DOE Office of Environmental Management is landlord until 2038 after which an unidentified federal agency will assume responsibility	Ongoing mission/no land use policy to date/central industrial area will be used for DP activities and environ. mgt.; end state for HLW tanks is scheduled for 2024	Various, depending on specific site; all land and groundwater located on site perimeter remediated for unrestricted use; institutional controls forever; groundwater strategy is a combination of ex situ and in situ treatments; soil contamination, buried waste, and buried structures will be contained by capping	2038

117

(*continued*)

TABLE B-1 Continued

State	Site	Responsibility for Site/End Use(s)	End State	Conditions of Closure	Completion Date
Tennessee	East Tennessee Technology Park (ETTP) (formerly K-25), ORR	Open space/recreational; controlled access; industrial with restrictions	Brownfield/part of site will be remediated to industrial levels as a private industrial park; part of site for restricted recreational use	Contaminated areas within re-industrialized area contained or consolidated; burial areas capped and hydrologically isolated and/or excavated; radioactive burial grounds will be capped; deed restrictions, monitoring and digging restrictions; groundwater monitoring until at least 2016	2013
Tennessee	Oak Ridge National Laboratory (ORNL), Oak Ridge Reservation (ORR)[4]	DOE Office of Science has ongoing mission	End states and their corresponding cleanup levels for the entire Oak Ridge Reservation are still being determined	Buried wastes isolated with engineered and institutional controls on migration; contaminated sediments stabilized; radioactive burial grounds will be capped and isolated; inactive bldg. razed to grade; stewardship will be needed for hundreds of years; federal gov't. will be responsible for site-wide groundwater monitoring and treatment forever	2013 (Stewardship until 2070 is planning basis)
Tennessee	Y-12, ORR	DOE Defense Programs has ongoing mission	Brownfield/controlled access/restricted industrial use; controlled access; open space/recreational use/ waste mgt. disposal facility for CERCLA waste	Burial ground contamination capped in place; groundwater contained and use restricted; stewardship will be needed for hundreds of years; federal government will be responsible for site-wide groundwater monitoring and treatment forever; site will maintain institutional controls and conduct CERCLA five-year reviews, inspections, monitoring and reporting; pump-and-treat systems may address on-site groundwater	2013 (Surveillance and monitoring of treatment systems through 2070)
Texas	Pantex Plant	DOE will keep control; site closure not expected in foreseeable future; current land use is called "industrial"	Brownfield/ongoing mission	Groundwater pump and treat may be required at least until 2015; identified release sites remediated to TX Risk Reduction Stds.	2002
Utah	Monticello Millsite and Vicinity Properties	Some land deeded to city for recreational use; on-site repository will remain under DOE control	Greenfield/controlled access	Remediation methods of sediments, groundwater, surface water not yet decided	2001

119

				2046
Washington	Hanford Site[5] 560-square-mile site near the Columbia River; includes the four major areas discussed below	Federal government maintains ownership of most of the site	Two site remediation goals: unrestricted and restricted use; specific land use decisions pending completion of Hanford Remedial Action EIS and Comprehensive Land Use Plan	DOE quarterly surveillance and routine radiological surveys; repair of barricades; vegetation management; surplus facilities D&D; compliance with tri-party agreements; institutional controls indefinitely to control groundwater use; semi-annual monitoring for at least 30 years after closing last facility
	100 Area: nine reactors		Restricted use	Institutional controls; reactors, N-fuel basin, K basins in interim safe storage for up to 75 years
	200 Area: reprocessing area		Federal government will use the area for management and disposal of nuclear materials; cleanup levels have not been established	Surface barriers over contaminated soil, waste sites and burial grounds; institutional controls; double shell and single shell tanks will remain; tanks will be closed in RCRA compliant manner; post-closure monitoring of tank farms through 2050; sanitary solid waste landfill will remain
	300/400 Area: fuel fabrication and support facilities		Final end state determined by ongoing CERCLA process	Radioactive cleanup to 14 mrem/year; deed restrictions used to control industrial use; institutional controls

Abbreviations Used in Table: CERCLA—Comprehensive Environmental Response, Compensation, and Liability Act of 1980, as amended; CH—contact handled; D&D—decontamination and decommissioning; DNAPL—dense nonaqueous phase liquid; EIS—Environmental Impact Statement; EPA—U.S. Environmental Protection Agency; FFACO—Federal Facilities Agreement and Consent Order; HLW—high-level waste; LLW—low-level waste; RCRA—Resource Conservation and Recovery Act of 1976, as amended; ROD—record of decision; RH—remote handled; TRU—transuranic; UMTRA—Uranium Mill Tailings Remedial Action; USEC—US Enrichment Corporation; DOE—U.S. Department of Energy.

1 Rocky Flats Environmental Technology Site is 6,185 acres. The majority of the land is uncontaminated and meets Residential land-use standards, but is currently limited to use as a buffer for plutonium stored on site. The core of the cleanup area is 384 acres, which will attain Industrial land-use standards to allow for environmental technology development activities. Determination of the future status of the Rocky Flats site is still very much a work in progress.

2 INEEL is the largest of the five major sites. Under the base case 99 percent of the area meets Residential use standard. Contaminated areas and facilities present only limited opportunities for alternative uses.

3 The SR site is a very complex site located in a humid environment. The majority of the surface of the site is uncontaminated. Contaminated surface waters and sediments limit the remainder of the site to Open Space use. The area north of the production area meets Agricultural use standard. The maximum feasible greenfield case for the SR site is limited by the possible end state for the five reactors, the chemical processing buildings and storage/disposal areas in the E, F, and H areas, which remain controlled access. Under some cleanup strategies most of the site could be brought to residential standards, however, most of the land will be used for resource or wildlife management.

4 The Oak Ridge site has a high water table. Although most of the site is uncontaminated the nature of the site and the three production areas limit use of that area to Open Space. A significant portion of the cost at the site is allocated to monitoring and addressing migration of contamination from numerous waste burial areas.

5 Most of the land currently meets residential standards. DOE actively uses only 8,150 acres for industrial/storage/disposal. Onsite plutonium storage is a major determinant of future land use because of buffer and emergency planning requirements. The base cleanup strategies assume complete dismantlement of reactors and core removal and extensive contaminated soil excavation. The 200 Areas remain Controlled Access for storage/disposal and waste mgt. activities.

APPENDIX C

Committee Information Gathering Meetings

The committee and subgroups of it conducted this study in part through meetings and site visits. The preponderance of meetings were for the purpose of gathering information through presentations by and discussions with representatives of the U.S. Department of Energy (DOE), its contractors, and other invited guests, followed by discussion among the committee members. All of the information gathering meetings were open to the public, and members of the public were given the opportunity to speak to the committee. A few meetings were closed to all but committee members and National Research Council staff; these sessions enabled free and critical discussion of findings and conclusions of the study as the committee prepared its reports. An additional means of communication used by the committee was conference calls among subgroup members (3 to 5 of the committee's 15 members) to plan site visits, clarify information, and coordinate preparation of reports for the full committee. The calls, having been numerous, are not included in the list below, which gives the dates and locations of the meetings, presentations received, and field trips taken.

February 27-28, 1997, Irvine, CA

Work Conducted by Resources for the Future (RFF) on Long-Term Stewardship at DOE Sites—James Werner, Director, Strategic Planning and Analysis, DOE Office of Environmental Management (EM-24)
Institutional Controls at DOE Nuclear Complex Sites—John Pendergrass, Senior Attorney at the Environmental Law Institute (ELI)

May 7-9, 1997, Washington, DC

National Research Council Studies Being Conducted for DOE/EM Concerning Technology Development—K.T. Thomas, National Research Council
Closure Issues at Oak Ridge National Laboratory, Oak Ridge, Tennessee—Margaret Wilson and Richard Ketele
Closure Issues at Hanford, Richland, Washington—Rich Holten
Closure Issues at Pinellas, Florida—Craig Scott
Closure Issues at the Nevada Test Site, Mercury, Nevada—Tom Longo
Work Conducted by Resources for the Future (RFF) on Long-Term Stewardship at DOE Sites—Kate Probst, RFF
U.S. Environmental Protection Agency (EPA) Policy on Closure and Institutional Controls—Ken Lovelace and Sharon Frey, EPA

APPENDIX C

July 28-29, 1997, Woods Hole, MA

Formerly Utilized Sites Remedial Action Program (FUSRAP) Managed for DOE by Oak Ridge National Laboratory—Al Johnson, DOE Office of Environmental Restoration (EM-40)
Closure and Post-Closure of Hazardous Waste Management Units under the Resource Conservation and Recovery Act (RCRA)—Travis Wagner, SAIC
Brownfield Cleanup Work in Chicago, Illinois—James Van der Kloot, EPA
Closure of Radioactive Waste Tanks at DOE Sites—Bob Bernero, retired from the U.S. Nuclear Regulatory Commission and a consultant to the committee

September 10-12, 1997, Las Vegas, NV, and Nevada Test Site, Mercury, NV

Introductions and Site Overviews—Terry Vaeth and Carl Gertz, DOE-Nevada
End State for the Nevada Test Site (NTS) and Soil Contamination and Remediation—Monica Sanchez, DOE-Nevada
Source Terms from Nuclear Test Events—Joe Thompson, Los Alamos National Laboratory
Source Terms from Waste Management Activities—Joseph Ginanni, DOE-Nevada
Environmental Monitoring—George McNeil and Robert Bangerter, DOE-Nevada
Groundwater Model Used at NTS to Predict Migration of Contaminants—Rick Waddell, HSI GeoTrans, Inc.
An Independent Site Risk Assessment—Don Baepler, University of Nevada, Nevada Risk Assessment Management Program (NRAMP)
Comments from the Public—Bob Loux, Executive Director of the Nevada State Nuclear Waste Project Office

The committee visited the Nevada Test Site (NTS) and the proposed high-level waste site at Yucca Mountain, Nevada.

November 4-5, 1997, Hanford Site, Richland, WA

Hanford Environmental Restoration Long-Range Plan—Rich Holten, DOE-Richland
Hanford Geology and Hydrology—Karl Fecht, Bechtel
Land Use Planning—Tom Fen, DOE Richland
Remediation Actions Planned for the Hanford 100 Area—Nancy Werdel, DOE-Richland
Burial Ground Strategy for the Hanford 200 Area—Jeff James and Brian Foley, Bechtel
Groundwater Protection Management Plan and the Monitoring and Analysis Program—Mike Thompson and Ron Smith, Pacific Northwest National Laboratory
Comments from the Public—Ralph Patt, Oregon State Department of Energy; Barbara Harper, representing Native Americans; and Jack Donnelly, Washington State Department of Ecology

The meeting was followed by a visit to the following facilities on the Hanford Site: 200 Area Hanford Barrier, 200 Area Environmental Restoration Disposal Facility (ERDF), and N Reactor.

March 4-6, 1998, Tucson, AZ

The committee, working with representatives from DOE and the Waste Management 1998 (WM'98) Program Planning Committee, organized and conducted the Closure and Institutional Controls Workshop at the WM'98.

April 14, 1998, Santa Barbara, CA

Two committee members met with Lorne G. Everett, Chief Research Hydrologist and Vice President, ARCADIS Geraghty & Miller, Inc., and Director of the Vadose Zone Monitoring Laboratory, University of

California at Santa Barbara. Subjects discussed included performance monitoring, performance demonstration, and interface modeling and monitoring for the vadose zone. The subgroup also visited Dr. Everett's vadose zone laboratory at the university.

April 20-22, 1998, Oak Ridge, TN

Before this meeting, several members participated in a half-day tour hosted by site representatives to observe locations that provide key background into the geological and hydrological features of the site. In addition, the members went on a field trip to several of the reservation facilities having relevance to near-term future uses and stewardship.

Site Closure Plans Such as Residual Hazards, Reindustrialization, the Watershed-Scale Approach to Cleanup Decisions, and Long-Term Water Use Restrictions—Rod Nelson, Bechtel/Jacobs

Near-Term Objectives, Future Land Use, and Long-Term Stewardship Requirements—Robert Sleeman and Charles Spoons, DOE Oak Ridge Operations (ORO)

Reindustrialization of Facilities at the East Tennessee Technology Park (ETTP; Formerly K-25) Through the Community Reuse Organization of East Tennessee (CROET)—Susan Cange, ORO

Data Available on Soil and Groundwater Contamination and Its Remediation—Ron Kirk, ORO

Burial Grounds at Bear Valley, and Hydrology and Closure and Post Closure Groundwater Monitoring for the Upper East Fork Poplar Creek—Karen Catlett and Margaret Wilson, ORO

Land Restriction—Certified Realty Specialists Mildred Ferve and Shirley Kates, ORO

Environmental Remediation at Oak Ridge National Laboratory (ORNL)—Ralph Skinner and Kavanough Mims, ORO

Comments from the Public—Bill Pardue, Chair, Site Specific Advisory Board; Al Brooks, Local Oversight Committee; Jim Phelps, Environment News and former employee at ORNL; Doug McCoy, Tennessee State Department of Environment and Conservation; James Hill, Local President, National Association for the Advancement of Colored People (NAACP); Susan Gawarecki, Executive Director of the Local Oversight Committee; and Cheryll Dyer, Coalition for a Healthy Environment

August 24-25, 1998, Grand Junction Project Office, Grand Junction, CO

Introductions—Jack Tillman, Director, Grand Junction Project Office (GJPO)
Grand Junction Project Office Responsibilities and Approach—Russel Edge, DOE GRPO
Records Management, Institutional Controls, and Site Performance Validation—Russel Edge, DOE GJPO

Following the meeting, members of the committee participated in a tour of the Cheney Disposal Cell and the Rifle, CO, closed UMTRA cell.

October 5, 1998, Mound Plant, Miamisburg, OH

The meeting was preceded by a walking and driving tour of the site.

Introductions and Background Information—Oba Vincent, DOE, Dick Neff, DOE Contractor, and Richard Church, Mayor of Miamisburg, OH
Mound 2000—Art Kleinrath, DOE Ohio
Transition Schedule—Sue Smiley, DOE Ohio
Post Remediation Control Systems—Randy Tormy, DOE
Future Plans—Dennis Bird, Miamisburg Mound Community Improvement Corporation (MMCIC)

APPENDIX C

October 5, 1998, Fernald Environmental Management Project, Ross, OH

Site Overview—Dennis Carr
Site Geology/Hydrogeology/Nature and Extent—Bill Hertel, Fluor Daniel
Remedy Selection/Setting of Cleanup Objectives—Marc Jewett, Fluor Daniel
Final Land Use/Natural Resources—Terry Hagen, Fluor Daniel
Fate and Transport Modeling/Statistics—J.D. Chiou, Fluor Daniel
Long-Term Monitoring—Mark Cherry, Fluor Daniel

The meeting was followed by a walking and driving tour of the site.

November 4-5, 1998, Washington, DC

Environmental Law Institute (ELI) Case Studies—Jim McElfish, ELI
Natural Attenuation/Intrinsic Remediation—Steve Golian, DOE/EM, and Jackie MacDonald, National Research Council

January 8, 1999, Grand Junction Project Office, Grand Junction, CO

A revisit to this office to discuss information management and record keeping.
Records Management, Institutional Controls, and Site Performance Validation—Russel Edge, DOE GJPO

April 28, 1999, Hanford Site, Richland, WA

Tour of the Hanford Site included the 100 Area burial grounds, B Reactor, K Basins, N Springs, old town sites, 200 Area tank farms, Environmental Restoration Disposal Facility (ERDF).

Composite Analysis—Charles Kincaid, Pacific Northwest National Lab
Vadose Zone Project—Mike Graham, Bechtel Hanford Inc., and Mike Thompson, DOE Richland
Hanford Remedial Action Environmental Impact Statement—Bill Edwards
Office of River Protection—Bill Edwards, DOE Richland
Paths to Closure and Stewardship—Jim Dailey, DOE Richland
Comments from the Public—Doug Sherwood, Region 10, U.S. Environmental Protection Agency; Dib Goswami, Washington State Department of Ecology; Barbara Harper, Yakama Nation

May 11-12, 1999, Augusta, GA, and Savannah River Site, SC

This meeting was preceded by a tour of the Savannah River Site, including stops at M-Area, Old F Seepage Basin, Burial Ground Complex, H Groundwater Treatment Unit, and Rainbow Bay—Jerry Nelson, Dean Hoffman.

Savannah River Operations Office (SROO) Welcome—Frank McCoy, Deputy Manager, SROO
Environmental Perspective—Tom Heenan, SROO
Land Use and Forest Services—Chris Noah, Chuck Borup, and Steve Stine, U.S. Forest Service
Environmental Monitoring—Bob Lorenz
Data Availability—Charles Murphy
Public Involvement—Mary Flora
Tank Closures—Larry Ling
Groundwater Modeling Overview—Mary Harris, Greg Flach
Savannah River Site Composite Analysis—Jim Cook, Elmer Wilhite
Technology—Sharon Robinson

Future Missions—Robert Meadors, SROO
Public Comment

June 9, 1999, Washington, DC

Industry Experience in Remediation, Institutional Controls, Regulatory Compliance, etc.—Edmund B. Frost, Attorney at Law, Leonard, Hurt, Frost & Lilly

November 15, 1999, Washington, DC

DOE/EM Long-Term Stewardship Program—James Werner and Andrew Duran, Office of Environmental Management.

APPENDIX D

Summary of Recent Stewardship Studies

Mary R. English

Over the past few years, a good deal of attention has been paid to the role of institutional controls and, more broadly, stewardship in arrangements for sites with potentially harmful residual contaminants. These sites include not only sites within the U.S. Department of Energy (DOE) nuclear weapons complex, but also other federal sites, such as those of the Department of Defense; Comprehensive Environmental Response, Compensation, and Liability Act of 1980, as amended (CERCLA, also known as Superfund) sites that are on the U.S. Environmental Protection Agency (EPA) National Priorities List; other sites within state Superfund programs; and sites being cleaned up under voluntary programs, leaking underground storage tank programs, and Resource Conservation and Recovery Act of 1976, as amended (RCRA) corrective action programs. In addition, the expected shut-down and decommissioning of a number of the nation's more than 100 nuclear power reactors has heightened attention to provisions for possible residual contaminants at their sites.

Most analyses assume that continued stewardship of some sites will be necessary; the question is not whether, but how. While these analyses vary in their degree of skepticism about the long-term workability of these arrangements, virtually none is altogether sanguine. Key findings and recommendations of some recent reports are summarized below.

Probst, K.N., and A.I. Lowe. 2000 (January). Cleaning Up the Nuclear Weapons Complex: Does Anybody Care? Center for Risk Management, Resources for the Future, Washington, D.C. 37 pp.

This study was funded by the W. Alton Jones Foundation, the John Merck Fund, and by general support from Resources for the Future. The DOE annual budget for its Office of Environmental Management (EM) is approximately $6 billion. Over $50 billion has been spent on cleanup to this time, and EM estimates that it will cost at least another $150 to $200 billion over the next 70 years to complete. Despite these costs, very little attention has been paid on the national level to DOE's cleanup program. Four of the many reasons for this lack of attention are: (1) it is difficult to focus on an environmental problem that is so large in scope and so technically complex; (2) most of the former weapons production facilities are in remote areas, far from population centers; and (3) funding and much of the oversight of DOE's environmental program fall under the defense committees in Congress, where even huge environmental outlays pale in comparison with other defense programs, and (4) DOE's EM program has become an important job-creation engine in the communities that once employed many in the nuclear weapons enterprise, making it a politically popular program.

The study concludes that increased national attention to and public scrutiny of the EM program at DOE is long overdue. DOE sites harbor contamination that will remain hazardous for thousands of years, and billions of dollars will be spent to reduce these risks in the coming decades. Mismanaged or misguided projects have cost taxpayers millions—perhaps even billions—of dollars in the past 10 years. Four steps are recommended to help improve the workings of the DOE environmental cleanup project: (1) clarify the EM mission and separate DOE's "job creation" and economic transition functions from EM programs and contracts; (2) decide which sites will—and which will not—have a future DOE mission; (3) require annual reports to Congress on the EM program; and (4) create an independent commission to evaluate the current EM organizational structure and identify needed reforms.

Stewardship Working Group. 1999 (December). The Oak Ridge Reservation Stakeholder Report on Stewardship: Volume 2. Oak Ridge Reservation End Use Working Group. Oak Ridge, Tenn.

This study was conducted at the request of DOE to the Oak Ridge Reservation Environmental Management Site Specific Advisory Board to describe the need for and the basic elements of a stewardship program, its application to contaminated areas on the DOE Oak Ridge Reservation, and the roles and responsibilities of stakeholders (defined as individuals, organizations, or other entities that have an interest in what happens to the Oak Ridge Reservation and other DOE facilities). The Stewardship Working Group recommended that:

(1) The Secretary of Energy issue a national policy establishing a commitment to long-term stewardship, to be followed by implementation guidance that allows for local participation and flexibility.

(2) DOE codify its approach to fulfilling its stewardship responsibilities in all CERCLA Records of Decision for the Reservation and in other legally binding documents. Interim Records of Decision must include project-specific stewardship requirements. Comprehensive long-term stewardship requirements for the Reservation must be described in final Records of Decision.

(3) The Federal Facility Agreement for the Reservation be amended to require and to develop appropriate milestones for the major stewardship-related documents, including the Reservation Land Use Control Assurance Plan, each project Land Use Control Implementation Plan, and final Records of Decision.

(4) DOE amend the Oak Ridge Reservation Public Involvement Plan and Federal Facility Agreement to provide for public and local government involvement in the following activities: the Reservation Land Use Control Assurance Plan; each project Land Use Control Implementation Plan; the DOE Long-Term Stewardship Plan; and five-year reviews.

(5) A Citizens Board for Stewardship be established or designated to review and assess long-term stewardship of the Reservation.

(6) DOE promptly recognize and work with all proposed stewards to begin implementation of their respective stewardship functions. The functions should be defined and incorporated into the DOE Long-Term Stewardship Plan.

(7) DOE implement, in cooperation with other entities, a Stewardship Research Program designed to understand the ecological and social impacts of residual contamination and to devise new and improved long-term remediation methods and technologies.

(8) DOE collect, preserve, and integrate all information needed for long-term stewardship of the Reservation in its information management system.

(9) DOE incorporate stewardship activities into a project management and tracking system to provide stewards with timely notification of stewardship activities and to track their progress.

(10) DOE implement a system of public information and education to disseminate timely information regarding environmental quality and required land use controls on the Reservation.

(11) DOE institute effective procedures for filing and registering contaminated land notices to ensure that they are found in title searches if land is transferred.

(12) DOE specify in relevant city, county, and state information systems the conditions and restrictions on the use of contaminated land.

(13) DOE continually refine its understanding of the specific costs of operating stewardship activities and incorporate these costs into the budget process.

(14) DOE identify for each remedial action the expected design life and the associated replacement or repair costs that can be expected by future generations.

(15) DOE, to the maximum feasible extent, promote mechanisms for funding stewardship that do not depend on annual appropriations, trust funds being the preferred approach. Should complete coverage of costs via trust funds not be possible, at least principal should be set aside to produce income sufficient for monitoring and other activities evaluating the impact of residual contamination on human health and the environment.

U.S. Department of Energy (Office of Environmental Management). 1999 (October). From Cleanup to Stewardship: A Companion Report to *Accelerating Cleanup: Paths to Closure* and Background Information to Support the Scoping Process Required for the 1998 PEIS Settlement Study. Office of Environmental Management DOE/EM-0466, Washington, D.C.

From Cleanup to Stewardship, produced by the Office of Strategic Planning and Analysis within the DOE Office of Environmental Management, is a companion report to *Accelerating Cleanup: Paths to Closure*. *From Cleanup to Stewardship* provides background information on the obligations and activities of long-term stewardship, which, according to the report, is expected to be necessary at more than 100 DOE sites after their remediation. The report discusses the nature of long-term stewardship at DOE sites, including such issues as activities to be performed, the regulatory context, the relationship of stewardship to land use, current organizational responsibilities, the largely unknown costs of long-term stewardship, and planning for long-term stewardship. The background information in the report supports the scoping process for a study required pursuant to the 1998 Lawsuit Settlement Agreement concerning the DOE Programmatic Environmental Impact Study (PEIS). The report includes the following five appendices: the December 1998 PEIS Lawsuit Settlement Agreement, Regulatory Requirements, Methodology, Glossary of Terms, and Site Profiles. The latter, which is contained in a separate document, provides brief analyses of 144 DOE sites that might need stewardship. Of these, 109 are, according to the report, expected to require some degree of long-term stewardship.

U.S. Department of Energy Environmental Management Advisory Board (EMAB) Long-Term Stewardship Committee. 1999 (September). Report and Findings for the September 1999 (EMAB) Meeting, Washington, D.C.

The Environmental Management Advisory Board (more commonly called EMAB) was developed to advise the DOE Office of Environmental Management (EM) on issues including site closure and stewardship. The Long-Term Stewardship Committee of EMAB identified four elements of an EM stewardship program: (1) site-specific stewardship plans; (2) a single, adequately staffed headquarters stewardship organization with responsibility for research, national decisions, and planning; (3) coordination with other DOE elements with stewardship needs; and (4) collecting and organizing relevant information and documenting current actions. The Committee recommended that EM adopt a national, written policy on long-term stewardship and detailed implementation guidance along with procedures to ensure a comprehensive analysis of all stewardship issues related to remediation decisions.

National Environmental Policy Institute. 1999 (September). Rolling Stewardship: Beyond Institutional Controls, Washington, D.C.

The goal of this report is to bring the issue of stewardship to the attention of key national, state, and local policy makers, particularly as it relates to public confidence in the reliability of present and future waste management strategies. The question the report poses to policy makers is what system of stewardship would be appropriate, given that sites requiring stewardship range from small, mildly contaminated brownfields to large and severely contaminated industrial and governmental sites. This in turn raises questions about the nature of a federal government role in stewardship, the roles of state, local, and tribal governments within stewardship and in relation to the federal government, and how stewardship should be funded.

English, M.R., and R. Inerfeld (Joint Institute for Energy and Environment, University of Tennessee, Knoxville, TN). 1999. Institutional Controls for Contaminated Sites: Help or Hazard? Risk: Health, Safety & Environment (Spring) 121-138.

The authors enumerate six requisite characteristics of institutional controls: effectiveness, appropriateness, verifiability, enforceability, durability, and flexibility. They then provide a critical review of four principal types of institutional controls: (1) deed restrictions based on common law, (2) deed restrictions based on state statutes, (3) local governmental land use controls such as zoning, and (4) other controls such as fencing, notification systems, and monitoring. They emphasize that deed restrictions should include explicit statements of intent, notice, and assignability and should, if possible, be grounded in statutory law; that local governmental land use controls should be a supplementary rather than a primary form of institutional control; and that institutional controls require oversight and enforcement conducted by entities that can be expected to remain in existence, with assignability of duties if they do not. The authors conclude that improvements in institutional controls are needed, because institutional controls are becoming an essential part of many remediation plans.

Environmental Law Institute. 1999. Protecting Public Health at Superfund Sites: Can Institutional Controls Meet the Challenge? Research Report, Environmental Law Institute, Washington, D.C. 121 pp.

The Environmental Law Institute (ELI) prepared this report with funding from the U.S. Environmental Protection Agency (EPA), but the views expressed are not necessarily those of EPA. The report investigates the effectiveness of institutional controls at four National Priorities List sites where there has been experience selecting and implementing various types of institutional controls: Cannons Engineering Corporation site in Bridgewater, Massachusetts, the Sharon Steel site in Midvale, Utah, the Cherokee Country site in Kansas, and the Oronogo-Duenweg Mining Belt site in Jasper County, Missouri. Some the findings include: (1) Institutional controls often depend on local government resources, authorities and agencies. (2) Cooperation among federal, state, and local governments in the implementation and operation of institutional controls is critical to their long-term efficacy. (3) Records of Decision (ROD) typically include only vague or general descriptions of institutional controls, although some RODs may be quite specific. (4) Institutional controls rely heavily on humans to implement, oversee, and administer them. (5) One method of reducing the risk of human error is to build redundancy into the institutional controls. (6) Institutional controls need to be monitored to assure that they are serving their intended purpose. (7) At some sites specific institutional controls are selected after the Record of Decision is signed. This takes these decisions out of the normal Superfund decision making process and, in particular, out of the normal process for public participation. (8) A fundamental element of the success of institutional controls is that community members to whom the controls apply understand their terms and the importance of compliance. (9) Records of Decision rarely include detailed information about how institutional controls will be implemented. (10) In addition to the direct costs of implementing institutional controls, their use can impose substantial indirect costs on communities, property owners, prospective purchasers and developers by limiting the way a site may be used.

Environmental Law Institute. 1999. Institutional Controls Case Study: Grand Junction. Research Report, Environmental Law Institute, Washington, D.C. 34 pp.

The Environmental Law Institute (ELI) prepared this report with funding from DOE, but the views expressed are not necessarily those of DOE. The study is a companion to the ELI 1998 case study on the DOE Mound Plant in Miamisburg, Ohio. Grand Junction, Colorado, was the site of processing of uranium ore that produced 2.2 million tons of tailings; the primary health risk associated with the tailings is exposure to radon gas, a radioactive decay product of radium that is naturally present in the tailings. Until the mid-1960s, the tailings were widely used as construction and fill materials, even within the City of Grand Junction. Institutional controls have been part of the Uranium Mill Tailings Remedial Action (UMTRA) program, including a database on sites and restrictions on use of groundwater. The report notes that the process of developing and implementing institutional controls has not been coordinated between DOE, the State of Colorado, the U.S. Nuclear Regulatory Commission, and the City of Grand Junction. The public generally has not been concerned about risks associated with the tailings, resulting in low public involvement in the UMTRA project.

State and Tribal Government Working Group Stewardship Committee. 1999 (February). Closure for the Seventh Generation.

Based on the deficiencies it found in the DOE current efforts to provide long-term protection of human health, the environment, and cultural resources, the committee made the following recommendations, among others. DOE should fully explain and quantify the long-term cost and funding required to maintain long-term institutional controls and it should acknowledge that decisions about long-term institutional controls will not be considered final until it has implemented an acceptable stewardship program. Each DOE site should develop a stewardship plan that defines constraints, costs, and implementation mechanisms. The DOE should retain ownership of lands at which institutional controls are necessary unless the appropriate state or tribe can certify that appropriate mechanisms are in place to enforce land use restrictions against subsequent users.

Applegate, J.S., and S. Dycus. 1998 (November). Institutional Controls or Emperor's Clothes? Long Term Stewardship of the Nuclear Weapons Complex. The Environmental Law Reporter ELR News & Analysis 28(11): 10631-10652.

The authors describe the CERCLA and RCRA framework affecting DOE waste management, the waste configuration options available to DOE (e.g., disposal facility, passive isolation in place, monitored retrievable storage), and the qualities of a long-term stewardship program to manage the DOE long-term wastes. Among these qualities are transparency (full and open assessment of risks), life-cycle accounting, documentation of nature and location of contaminants, identification of stewards, enforceability, redundancy, public involvement, sustainability, and flexibility and responsiveness. An institution capable of providing long-term stewardship of wastes will need to perform a generally continuous set of functions, raise funds to sustain those functions, transmit knowledge of itself and its values to subsequent generations, and maintain a core identity while adapting to changing circumstances.

U.S. Department of Energy. 1998 (June). Proceedings of Long-Term Stewardship Workshop. Grand Junction Office CONF-980652, Grand Junction, Colo. 198 pp.

The DOE Grand Junction Office held a workshop on long-term stewardship on June 2-3, 1998, in Grand Junction, CO. The stated goal of the workshop was to share ideas and evaluate solutions to problems associated with long-term custodianship of radioactive waste disposal sites. The proceedings of the workshop included eighteen papers, grouped into the following categories: regulatory perspectives, project management and records management, technical issues, performance monitoring and stakeholder issues, and site transfer protocols. Approximately half of the authors are affiliated with DOE or DOE contractors; the remainder are affiliated with state or federal regulators or with organizations in Germany, Australia, and Estonia.

Environmental Law Institute. 1998. Institutional Controls Case Study: Mound Plant. Research Report, Environmental Law Institute, Washington, D.C. 40 pp.

The Environmental Law Institute (ELI) prepared this report with funding from DOE, but the views expressed are not necessarily those of DOE. To help inform decision-making across the DOE complex, ELI studied the ongoing decision-making process—particularly the use of institutional controls as a component of cleanup and reuse—at the DOE Mound Plant in Miamisburg, Ohio. (In 1994, DOE decided to end defense production activities at the site. The site is intended for industrial reuse, and reuse has begun on portions of it.) The authors observe that issues have arisen concerning the enforceability of some of the site's deed restrictions, including what entities would be able to enforce them and the effects of the restrictions on marketing the site for reuse. They note that some institutional controls must be developed early in the transition to reuse, particularly if the controls rely on property concepts. They also note that consideration should be given to other institutional controls as a substitute or supplement to deed restrictions.

International City/County Management Association. 1998 (April). Local Governmental Use of Institutional Controls at Contaminated Sites. Washington, D.C.

This report is based on a survey of 27 local government officials as well as five interviews with state officials. The report concludes by identifying six recommendations in areas that, according to the report, appear to need the most improvement: (1) minimize reliance on institutional memory, (2) clarify jurisdictional issues and improve coordination between state and local governments, (3) provide training and education to local governments regarding the role and use of institutional controls, (4) improve the quality and increase the use of mechanisms for recording institutional controls, (5) improve the longevity of institutional controls, and (6) improve and increase enforcement efforts by providing adequate code enforcement resources.

ICF Kaiser Consulting Group. 1998 (March). Managing Data for Long-Term Stewardship, Working Draft. Washington, D.C.

This report was prepared for the DOE Office of Strategic Planning and Analysis as an information resource but does not represent official DOE policy or guidance. It is concluded in the report that, while most types of information needed for "long-term stewardship" are already being generated, there are a number of problems. For example, requirements do not specify what constitutes stewardship data, and information management requirements are not coordinated with property transfer requirements; some data will not be preserved as long as necessary for stewardship purposes (in fact, most records of facilities and site infrastructure are required to be destroyed when facilities are demolished or infrastructure is declared obsolete); some data will be preserved adequately but may be forgotten, not be easily located, or accompanied with insufficient descriptive information to be usable; and even when knowledge is preserved and users know where information is located, it may take too long or be too expensive to gain access to stewardship data.

Environmental Law Institute. 1998 (March 2). Preliminary Memorandum: Institutional Controls over Land Uses at Superfund Sites—Draft. Washington, D.C. 17 pp.

This memorandum was prepared under a cooperative agreement with the U.S. Environmental Protection Agency but does not represent official EPA views. In the memorandum, evaluations based upon hypothetical (not actual case studies) are provided of the effectiveness of regulatory controls (zoning, groundwater withdrawals); property-based controls (covenants, easements); and government-supplied notice. It is concluded that each of these types of institutional control has both strengths and weaknesses, and that in practice, the efficacy of a particular control will vary from site to site depending upon numerous factors.

U.S. Environmental Protection Agency. 1998 (March). Institutional Controls: A Reference Manual, Workgroup Draft. Workgroup on Institutional Controls. Washington, D.C.

This manual was prepared by U.S. EPA staff, including various attorneys and others in headquarters and regional offices. The manual is intended as an aid, not as policy guidance. It identifies various legal and other vehicles that can serve as institutional controls and discusses legal and practical considerations that may arise in putting such controls in place. In addition, it contains recommendations that the workgroup believes can improve future efforts to evaluate and implement institutional controls. Key recommendations are as follows: (1) institutional controls should be evaluated carefully before the remedy is chosen, (2) goals of institutional controls should be described clearly in the remedial decision document, (3) state and local agencies have a vital role in developing, establishing and maintaining effective and enforceable institutional controls, (4) misconceptions about the effect of simply specifying use restrictions in a deed should be corrected, (5) the term "deed restriction" should be used carefully, (6) the limitations of deed notices should be clearly understood, and (7) where it is important for the control to run with the land, it is generally not advisable to rely on a consent decree alone, without execution of a separate instrument such as an easement.

Probst, K.N., and M.H. McGovern. 1998 (June). Long-Term Stewardship and the Nuclear Weapons Complex: The Challenge Ahead. Center for Risk Management, Resources for the Future, Washington, D.C. 67 pp.

The research for this report was funded in large part by the DOE Office of Environmental Management, but the views expressed do not represent official DOE policy. The report addresses the question of what is needed in terms of a "long-term stewardship" program for DOE former weapons production sites. It concludes that the primary locus for stewardship should be the federal government, with involvement of other entities such as states, localities, tribal nations, and the general public. It recommends that (1) Congress should legislatively require the creation of a stewardship program for all contaminated sites requiring "post-closure care" that are regulated under the nation's environmental laws; (2) EPA should, in its regulations, clearly define post-closure responsibilities at Superfund sites on the part of federal, state, and local governments, and regulated entities; (3) the President's Council on Environmental Quality, with EPA, should convene an interagency task force to develop government-wide policy on long-term stewardship at both federal and private sites; and (4) the Secretary of Energy should create a high-level, diverse task force to develop a stewardship mission for DOE, and to make specific recommendations for integrating the costs and challenges of long-term stewardship into the major DOE decision and budgeting processes.

English, M.R., D.L. Feldman, R. Inerfeld, and J. Lumley. 1997 (July). Institutional Controls at Superfund Sites: A Preliminary Assessment of their Efficacy and Public Acceptability. Joint Institute for Energy and Environment, University of Tennessee, Knoxville, Tenn. 100 pp.

This report was prepared under a cooperative agreement with the U.S. Environmental Protection Agency's Office of Policy, Planning, and Evaluation but does not represent official EPA policy or guidance. The report focused on an evaluation of drivers for the public acceptability or non-acceptability of institutional controls at Superfund sites but included a preliminary assessment of the efficacy of institutional controls, based upon analyses in the legal and other literature. In the report, it was concluded that, based upon the small sample of cases studied, public attitudes toward institutional controls did not appear to be a significant barrier in most cases; however, there is reason for attention to and further research on the long-term efficacy of many institutional control measures.

Hersh, R., K. Probst, K. Wernstedt, and J. Mazurek. 1997 (June). Linking Land Use and Superfund Cleanups: Uncharted Territory. Center for Risk Management, Resources for the Future, Washington, D.C. 107 pp.

This research for this report was supported in part by the U.S. Environmental Protection Agency's Office of Emergency and Remedial Response and Office of Policy Analysis, but the views expressed do not represent official EPA policy. The report addresses the concept of linking land use to remedy selection in the Superfund program. It sets forth the following key findings and recommendations: (1) agreement about the future use of a site may not lead to agreement about the appropriate remedy or cleanup standards for that site, (2) it is often not possible to determine the "anticipated future use" of a site, and, in fact, the remedy selection process can lead to unanticipated land uses, (3) institutional controls are often critical to ensuring long-term protection; often neglected and left to the end of the remedy selection process; and subject to legal, administrative, and social pressures that may limit their effectiveness, (4) linking cleanup decisions to land use considerations places an even heavier responsibility on EPA to effectively involve the public in the remedy selection process, (5) EPA should revise its regulations to address the role of land use in remedy selection, including incorporating the development of institutional controls into the formal remedy selection process, and (6) in consultation with state and local governments, EPA should develop a strategy (ultimately codified in the National Contingency Plan) for ensuring effective long-term regulatory oversight of Superfund sites where contamination remains at levels that present a risk to public health even after the remedy is "complete."

International City/County Management Association. 1996. Cleaning Up After the Cold War: The Role of Local Governments in the Environmental Cleanup and Reuse of Federal Facilities. Research and Development Department, Washington, D.C. 123 pp.

This study was conducted under a cooperative agreement with the U.S. Environmental Protection Agency but does not necessarily represent the views of EPA. Using 11 case studies (seven Department of Defense sites and four DOE sites) as a basis, the study explores the roles and responsibilities of local governments in communities that host federal facilities with defense missions. Specifically, the authors analyze the roles and responsibilities of local governments in site cleanup and future use decisions. The authors provide a number of recommendations to facilitate more effective involvement of local governments in these decisions. The focus of these recommendations, on the one hand, is the need for federal officials to acknowledge that local governments have an important role to play and should be kept informed and involved, and, on the other hand, is the need for local governments to assume a more active and authoritative role in decisions concerning future land use restrictions, as well as in monitoring and oversight activities.

APPENDIX E

Existing Legal Structure for Closure of the Weapons Complex Sites

Elizabeth K. Hocking, W. Hugh O'Riordan, and *Robert M. Bernero*

There is no single federal or state legal framework governing closure of all DOE weapons complex facilities. In closing a facility, the U.S. Department of Energy (DOE) must comply with a panoply of environmental, radiological, and land use laws implemented by federal, state and local governmental authorities. Closure is characterized by wide variability, thus requiring diversity and flexibility from site to site.

The department's legal interest in and control over its land are not necessarily the same nationwide. In western states, arcane federal public land statutes are potent and may affect closure decisions. In eastern states, DOE land was purchased primarily and the nature and extent of the acquired property rights may not often be immediately clear. Similarly, state land laws can be critical to long-term closure decisions.

DIVERSITY OF DOE CLOSURE PROCESS

Every DOE site is subject to a varying mix of federal, state and local laws and regulations. In spite of the apparent rigidity of the very complicated applicable statutes and regulations, sites apparently negotiate a flexible system designed to meet their unique circumstances. In some cases, state law appears to dominate, where in others federal laws govern the process.

Not every site is moving toward the same concept of closure and stewardship. There are several reasons for this. The governing statutes are not consistent in how they establish directions and obligations for stewardship. The leadership for closure appears to be vested in the DOE field managers; DOE Headquarters direction and coordination do not result in uniformity. There is also little intersite comparison and interaction on similar closure situations. These factors result in no single pathway for the closure process.

This lack of a single closure pathway appears to have been accepted by all parties. All parties seem to be seeking the flexibility necessary to help them achieve their goals. For example, the Hanford Site is being cleaned up pursuant to an elaborate and complex Resource Conservation and Recovery Act of 1976, as amended (RCRA)/Comprehensive Environmental Response, Compensation, and Liability Act of 1980, as amended (CERCLA—also know as Superfund) Tri-Party Agreement (TPA) among the State of Washington, the U.S. Environmental Protection Agency (EPA), and DOE. The Nevada Test Site is being cleaned up pursuant to state of Nevada and DOE agreements under the auspices of RCRA. Several tanks at the Savannah River Site were remediated under the Clean Water Act. This range may be a case of regulators relying upon the legal framework they know best or the one they perceive to give them the most authority and flexibility.

CURRENT FEDERAL AND STATE ENVIRONMENTAL LAWS

Most DOE sites will be subject to either the federal RCRA or CERCLA, or both. Both statutes are of recent origin, and both were amended to clarify that they do apply to federal facilities (RCRA in 1992 through the Federal Facilities Compliance Act and CERCLA in 1986 through the Superfund Amendments and Reauthorization Act). The laws are written to provide general control of situations where hazardous substances on a site require some form of management and remediation. Both laws will likely change in the future.

The significance of state authority over remediation of DOE facilities within their borders cannot be underestimated. Many states control corrective action programs through their EPA authorized RCRA programs and environmental restoration through their own CERCLA analogues. Federal facilities not on the National Priorities List (NPL) are subject to state laws on remediation and removal actions (CERCLA Section 120[a][4]). Congress also provided states the opportunity in CERCLA Section 120(e) to participate in the development of remedial investigations and feasibility studies with DOE and EPA at sites on the NPL. Notice must be given to the affected state within six months of a federal facility being placed on the NPL.

In addition to these environmental remediation laws, DOE sites are subject to other older (25 years) major federal environmental statutes such as the Clean Water Act, Clean Air Act, Toxic Substance Control Act, Endangered Species Act, and National Environmental Policy Act. Many states will also have laws patterned after these federal laws, and some provisions of the states' laws may be more stringent than the federal laws. Each of these statutes has it own significant regulatory framework and standards that can become site-specific cleanup levels.

Federal Radiation Laws

In contrast to the general style of the environmental remediation laws, the "atomic laws" distinguish the type of radioactive material to be controlled and set specific requirements for such controls, including different site closure and custody requirements. Thus, DOE stewardship activities will vary depending upon the type(s) of material requiring management (e.g., low-level, transuranic, high-level, or mixed wastes).

Atomic Energy Act

The federal regulation of radioactive materials is based principally on the Atomic Energy Acts of 1948 and 1954, as amended. These acts place radioactive materials and practices under the authority of the Atomic Energy Commission, but only if they are associated with the nuclear fission process. Radioactive materials and practices not associated with the nuclear fission process are subject to regulation by the EPA and the states under their public health responsibilities.

The Energy Reorganization Act of 1974 divided the Atomic Energy Commission into the Nuclear Regulatory Commission (USNRC) and what is now the DOE. Existing federal regulatory responsibilities were generally split along the lines of the USNRC regulating commercial uses and DOE regulating itself in nuclear weapons and material research and management programs. Subsequent laws addressed specific aspects of the federal and state regulatory program, assigning specific responsibilities to DOE, USNRC, and state governments.

Uranium Mill Tailings Radiation Control Act of 1978 (UMTRCA)

The UMTRCA was established to regulate uranium mill tailings and the contamination associated with the mill sites. Although uranium and thorium are naturally occurring radioactive materials, they are covered by federal regulation since they are source material, a source of fuel for the nuclear fission process. The processing of source material evolved from closely guarded batch operations during World War II, typically in the eastern United States, to the operation of large uranium mills, typically in the western part of the country, as the need grew. The tailings from these mill operations are enormous piles, and emit radon gas as a daughter product of natural radioactive decay, with the amount of this short-lived gas sensitive to whether the pile is covered or not.

The UMTRCA was established in two titles. Title I addressed abandoned or orphaned mill sites and Title II addressed those commercial uranium facilities with current USNRC licenses at the time of the legislation. Congress mandated that EPA promulgate regulations dealing with the cleanup and remediation of contamination associated with the mill sites and that the USNRC enforce the EPA regulations. The Title I facilities were assigned to the DOE for cleanup and remediation. Remediation of the Title II facilities was the responsibility of the individual license holders. Upon successful remediation of the Title II sites, they were to be transferred to the federal government for long-term care. In establishing the legislation, Congress mandated that reliance on active remediation be minimized.

The EPA established regulations governing uranium mill tailings in 40 Code of Federal Regulations (CFR) Part 192. These regulations require, among other things, a disposal cell design life of 200 to 1,000 years, release limits for the tailings covers, and establishment of groundwater protection standards for each site. The EPA also promulgated groundwater clean up standards for the mill tailings sites. The principal result of these regulations is that tailings are to be emplaced in stable, capped piles with controlled releases of gases and water-leached materials.

In response to the EPA regulations, the USNRC established 10 CFR Part 40, Appendix A, which sets the technical criteria for tailings disposal. The USNRC also established 10 CFR Sections 40.27 and 40.28 to license DOE for custody and long-term care of Title I and Title II sites. The USNRC established a variety of guidelines for design of tailings disposal cells, including requiring the disposal cells to withstand a variety of natural forces, such as probable maximum floods and maximum credible earthquakes. Operations under these regulations are widespread, and the DOE-required custody is being exercised by the Grand Junction Operations Office for these uranium mill tailing sites, along with several other sites.

Low-Level Radioactive Waste Policy Act of 1980 (LLRWPA)

The LLRWPA set the initial framework for state responsibilities in the management and disposal of commercial low-level radioactive waste. Low-level radioactive wastes generated or stored at DOE sites remain the responsibility of DOE. The 1985 Amendments to the LLRWPA made clear that the disposal of Greater-Than-Class-C (GTCC) waste is the responsibility of DOE.

Nuclear Waste Policy Act of 1982 (NWPA)

The NWPA assigned DOE the responsibility to select and develop sites for the deep geologic disposal of high-level radioactive waste (HLW). The DOE is also responsible for indefinite custody and institutional management of such repositories. These sites are to be licensed by the USNRC to meet applicable EPA standards. The definition of high-level radioactive waste was expanded from the solvent extraction wastes of spent reactor fuel reprocessing to include the spent reactor fuel itself (since the United States has stopped spent fuel reprocessing). Through a presidential memorandum issued not long after the 1982 Act, President Reagan directed the use of commercial repositories for disposal of the DOE defense high-level waste (Memorandum from President Ronald Reagan to John S. Harrington, Secretary of Energy, dated April 30, 1985).

Nuclear Waste Policy Amendments Act of 1987 (NWPA)

This law reduced the burden on DOE of characterizing three high-level-waste deep-geologic-disposal sites simultaneously to characterizing only one at a time. DOE was directed to start with the Yucca Mountain Site in Nevada. The USNRC remained as the licensing body for the site and the EPA standard continued as a requirement.

Land Withdrawal Act of 1991 (LWA)

The LWA withdrew the needed federal lands near Carlsbad, NM, for the Waste Isolation Pilot Plant (WIPP), a deep repository for permanent disposal of defense transuranic waste, waste subject to the same EPA disposal standard as high-level waste. The act incorporated a unique regulatory requirement. It recognized that, under its

Atomic Energy Act authority, DOE had regulatory oversight and authorization of its own actions to develop and operate WIPP. The act stipulated, however, that EPA review and concur with DOE action.

Energy Policy Act of 1992

The Energy Policy Act of 1992 made a number of changes that are important to DOE waste disposal efforts. It rejected the EPA disposal standard for Yucca Mountain and directed EPA to seek a study by the National Academy of Sciences for recommendations on preparing a health-based waste disposal standard. That advice was provided in the National Research Council (1995) report, *Technical Basis for Yucca Mountain Standards*, followed by a letter report to EPA commenting on the report *Environmental Radiation Protection Standards for Yucca Mountain, Nevada; Proposed Rule* (64 Federal Register 46976-47016, August 27, 1999) (National Research Council, 1999). After hearing that advice, EPA is to promulgate a Yucca Mountain standard and the USNRC is to issue its revised licensing standard for the site. The 1992 Act also established the now private United States Enrichment Corporation to operate the DOE-owned gaseous diffusion plants in Kentucky and Ohio under lease. The DOE retains ownership and the responsibility to decommission and close the sites later.

Federal Radiation Laws and Stewardship Decisions

The radiation laws make DOE responsible for its own radioactive wastes, UMTRCA sites, deep disposal of high-level wastes generated by nuclear power plants, and wastes generated by the U.S. Enrichment Corporation, and potentially responsible for state compact disposal sites. The laws apply conditions to DOE management of these wastes but DOE possesses a great deal of authority to self-regulate its activities.

There has been recent discussion, and some limited congressional action, pointing toward imposing external regulation of DOE nuclear activities by the USNRC. However, there is little evidence of this as a growing trend in the site closure arena. The USNRC oversight is generally limited to statutory responsibilities, such as oversight of uranium mill tailings disposal, and some consultation, and review and comment on the DOE determination of incidental waste at the Savannah River waste tank farms. As a practical matter DOE is able to determine what it wants to do concerning site closure processes within the framework of the statutes and its own waste management standards. However, the DOE waste classifications and oversight role of several organizations could constrain DOE consideration of waste management alternatives that can impact stewardship decisions.

DOE Waste Classification and Disposal Requirements.

In accordance with DOE Order 435.1 Radioactive Waste Management, the DOE operates with waste management and disposal requirements associated with waste classes, which substantially affect plans and options for management of buried and tank wastes. For example:

• high-level waste (HLW) is defined by origin, from first state solvent extraction in fuel reprocessing or equivalent;
• spent fuel is classified as spent fuel, not waste, until a decision is made on its use or disposition as high-level waste;
• transuranic waste (TRU) is defined by transuranic concentration;
• low-level waste (LLW) is defined by exclusion (i.e., that which is not HLW, TRU, etc.); and
• the presence of RCRA hazardous material can make the radioactive material mixed waste, requiring additional treatment prior to disposal.

Buried or tank waste at a DOE site, when it is characterized, is considered for disinterment or extraction in order to be managed in the proper waste stream. The preferred DOE path is to leave only radioactive residues to be considered for release of the site. In some circumstances, practical matters force the consideration of higher level residues, making the site in question a radioactive waste disposal site, with the need to satisfy the requirements for the type of waste being disposed. This is the approach now being taken with the liquid waste tanks at Savannah River once they are emptied. At the Nevada Test Site, removal of the waste from the detonation cavities is not practical, so other approaches must be considered.

External Oversight and the Effects of Change

Although the DOE has the authority to self-regulate its radiation activities, it does operate with a substantial degree of external oversight. Many facilities are subject to oversight by the Defense Nuclear Facilities Safety Board (DNFSB).[1] The DNFSB does not perform any licensing or permitting of DOE facilities, but does provide structured review of DOE practices and formal advice to DOE.

The USNRC already has statutory responsibility to regulate DOE in storage and disposal of HLW under the Nuclear Waste Policy Act of 1982 and the Nuclear Waste Policy Amendments Act of 1987. The DOE has the responsibility to receive and dispose of, with USNRC approval, the subset of commercial LLW that is Greater-Than-Class-C (GTCC), that is, LLW that contains radioactivity in excess of the limits set for Class C LLW.

The USNRC has been engaged in substantial interaction with DOE to provide advice and concurrence with DOE plans for the management of liquid wastes that are potentially HLW (Bernero, 1993; Paperiello, 1997). The liquid waste management criteria emerging from this interaction are quite simple: technically and economically practicable extraction of the waste and concentration into HLW, and treatment of the low-activity extract and residues to the performance standards for LLW.

DOE Closure Standards

As indicated earlier, DOE has established an extensive array of orders and guidance documents to guide its activities. The DOE recently released Order 435.1, Radioactive Waste Management, and an accompanying guidance document, Radioactive Waste Management Manual. The Manual clearly defines the closure procedures and requirements applicable to DOE management of radioactive wastes.

The closure of high-level waste facilities and sites can be accomplished through (a) decommissioning to the point that the facility or site can meet the release restrictions in DOE Order 5400.5 (Radiation Protection of the Public and the Environment), (b) deactivation in accordance with CERCLA, or (c) development and implementation of an approved closure plan. The closure plan must address operational or interim closure, final facility closure, and institutional closure. Operational/interim closure and final closure consist of the physical activities preparatory to facility closure. Institutional closure occurs after final closure and consists of all actions and measures necessary to ensure the long-term stability of the site. The closure plan must include a monitoring plan that includes the location of monitoring wells or monitoring points, the data to be collected, and the actions that will be taken in response to the results of the monitoring data. The closure plan must also include land use limitations and other institutional controls that must be in place until the facility or site can be released for unrestricted use.

Plans are also required for all low-level waste disposal facility closures. The preliminary closure plan is submitted with the facility performance assessments and composite analyses and is updated throughout the operational life of the facility. Upon closure, institutional control measures are to be integrated into land use and stewardship plans and programs and must be continued until the facility can be released according to the requirements of DOE Order 5400.5.

For both high-level waste facilities and sites and low-level waste disposal sites, a 100-year period of active institutional controls is normally assumed. During this period, access would be controlled and monitoring and custodial maintenance would be performed. Longer periods of active institutional controls may be assumed if justified in documented plans.

---

[1] DNFSB was established by the Congress to provide independent safety oversight on DOE defense nuclear activities that are self-regulated by DOE under its own Atomic Energy Act authority.

Federal and State Land Laws

Federal and state land laws are the oldest legal frameworks affecting closure of DOE facilities. Neither the federal public land laws nor state land laws were designed specifically for closure of weapons facilities. Moreover, both federal public land laws and state land laws view ownership and control of land as a bundle of rights. Water rights, mineral rights, and rights of way can be held independently of "ownership" of land. This means that the DOE rights as "landowner" may vary from site to site.

Federal and state land laws are very important for the closure process. If DOE residually contaminated land is going to be leased, sold, or granted, some form of institutional control will probably be needed to enforce a use restriction. Many of the typically conceived institutional controls—deed restrictions, easements, zoning, construction or excavation permits, and groundwater use restrictions—are dependent upon the authorities found in state, local, or tribal law.

Federal public land laws prescribe the use and disposition of DOE land. Approximately 62 percent of DOE land was withdrawn from the United States public domain lands by the Department of the Interior (DOI) for specific DOE mission purposes. These "withdrawn" lands must be relinquished to the DOI when they are excess to DOE. Real property is excess when it is no longer required for DOE needs and the discharge of Department responsibilities (DOE Order 430.1A, Life-Cycle Asset Management). If the DOI does not want the land back—because it is so substantially changed in character that it is not suitable for public land—the land is returned to DOE for holding or disposition through other means.

Approximately 27 percent of DOE land was acquired from private owners by the federal government for use by DOE. When these acquired lands are excess to DOE, the Department can lease, sell, or grant them if the new use comports with the purposes of the Atomic Energy Act, Section 161(g). The DOE can also lease excess acquired land under the Hall Amendment to the DOE Organization Act. Alternatively, the Department can turn these excess acquired lands over to the General Services Administration for disposition. The Atomic Energy Act and the DOE Organization Act also allow for the leasing of DOE lands that are temporarily not needed by, but not yet excess to, the Department.

The FY 1998 National Defense Authorization Act requires the DOE to prescribe regulations for its transfer of real property at its defense nuclear facilities for economic development of the property. The Department is in the process of writing those regulations. Disposition of DOE facilities may also be affected by tribal claims to federal land. Claims may be based on treaty rights, the trust responsibility between the federal government and the tribes, aboriginal lands, or ceded lands.

The Public's Role in Closure

Members of the public use the existing legal structure to influence DOE cleanup and closure efforts. Environmental laws and guidance documents strongly encourage public involvement in determining future land uses, cleanup levels, and remedies.

CLEANUP LEVELS AND INSTITUTIONAL CONTROLS

Some flexibility is built into the legal framework for establishing cleanup levels based on future land use and use of institutional controls. While CERCLA gives a preference for treatment and remedy permanence (i.e., complete cleanup), it and RCRA allow hazardous wastes to remain on site. The agreed-upon future land use is a key to determining cleanup levels (Land Use in the CERCLA Remedy Selection Process, EPA/OSWER 9355.7-04, May 25, 1995) and institutional controls can be elements of a CERCLA remedy (National Contingency Plan, 40 CFR 300.430[a][1][iii][D]).

Cleanup Levels and Future Land Use

Site cleanup levels are driven by a combination of site risks, future site use, and federal and state requirements applicable to the site. Risks posed by a site include risks to workers, the public, and the environment before, during, and after remediation. Communities surrounding sites and the states hosting them play an important role in determining site future uses. Future uses can include, for example, residential, grazing, recreational, wildlife refuge, commercial, or industrial uses.

Actual cleanup levels reflect the expected future land use and are drawn from standards, requirements, criteria, or limitations established in federal or state environmental laws. In some cases, complete cleanup is not technically practicable due to the nature of the source and existing technologies (e.g., the treatment/control of dense non-aqueous phase liquid [DNAPL] contamination of groundwater) or is not economically feasible.

If contaminants remain on the site, use restrictions will be put in place to ensure that people and the environment are not unduly exposed to the contaminants. Use restrictions in the form of institutional controls could include, for example, limited site use of no more than eight hours a day, well-water use bans, and excavation limitations.

Institutional Controls Under CERCLA and RCRA

Institutional controls can be used under CERCLA "to supplement engineering controls as appropriate for short- and long-term management to prevent or limit exposure to hazardous substances, pollutants, or contaminants" (40 CFR 430). However, according to CERCLA regulations, institutional controls should not be used as the sole remedy unless active response measures are not practicable. The decision about the impracticability of active response measures requires considering remedy selection factors such as the overall protection of human health and the environment; compliance with applicable or relevant and appropriate requirement (ARAR); long-term effectiveness and permanence; reduction of toxicity, mobility, or volume through treatment; short-term effectiveness; implementability; cost; and state and community acceptance (40 CFR 300.430[a][1][iii] [D]).

Whenever hazardous substances, contaminants, or pollutants remain on site as part of the CERCLA remedy, the remedial action must be reviewed "no less often than each 5 years" to ensure that the remedial action is protective of human health and the environment (CERCLA, Section 121[c]; Executive Order 12580). Institutional controls would also be reviewed under this requirement. The remedial action review requirements for RCRA remedies would be listed in the facility permit.

Two EPA regions, IV and X, have issued policies on the use of institutional controls at federal facilities. Both regions require federal facilities using institutional controls to submit plans explaining how the effectiveness of the institutional controls will be ensured through time.

Land Transfers of Contaminated Property

CERCLA

Under CERCLA, deeds transferring federal property on which any hazardous substance was stored for one year or more, or known to have been released or disposed, must include the following information to the extent it is available from a complete review of agency files:

1. notice of the type and quantity of such hazardous substances;
2. notice of the time at which the storage, release, or disposal took place; and
3. a description of any remedial action that was taken.

Deeds transferring property must also contain covenants warranting that (1) all remedial action necessary to protect human health and the environment has been taken before the property is transferred, and (2) any additional remedial action necessary after the transfer will be conducted by the federal government (CERCLA Section

120[h][3][A]). For purposes of the first covenant, the statement that all remedial action has been taken means that the construction and installation of an approved remedial design is completed and the remedy has been demonstrated to the EPA as operating properly and successfully (CERCLA Section 120[h][3][B]).

Federal property can also be transferred under CERCLA even if remedial action has been deferred when the EPA or state governor, as appropriate, determines that the property is suitable for the intended use, the intended use is consistent with protection of human health and the environment, and the deferral of remediation and the property transfer will not substantially delay necessary response actions. The public in the general vicinity of the property must be given a chance to comment on the transfer. When remedial action is deferred, the federal agency must warrant in the deed or property transfer document that it will provide (CERCLA Section 120[h][3][C]):

1. any necessary restrictions on the use of the property to ensure protection of human health and the environment;
2. use restrictions to ensure that required remedial investigations, response actions, and oversight activities will not be disrupted;
3. necessary response actions and schedules for investigations and completion of response actions; and
4. budget requests adequate to cover response actions to the Office of Management and Budget.

RCRA

When federal agencies close a unit under RCRA, they must submit a survey plat indicating the location and dimensions of landfill cells or other hazardous waste disposal units with respect to permanently surveyed benchmarks to the local zoning authority or the authority with jurisdiction over the site (40 CFR 264.119). The federal agency must also, in accordance with the applicable state procedure, record a notation on the deed to the facility property—or any other instruments that would normally be examined during a title search—that will "in perpetuity" notify potential purchasers that the property had been used to manage hazardous wastes, that its use is restricted to maintain remedy integrity, and that a survey plat has been filed (40 CFR 264.119).

NEPA Documentation

Federal agencies must complete appropriate documentation under the National Environmental Policy Act (NEPA) when transferring land. A categorical exclusion (CX) from the NEPA requirements may be appropriate if the impacts of the post-transfer land use would remain essentially the same as the pre-transfer impacts and there are no intervening variables that could cause significant environmental issues. If the expected land use will be a change in usage, either an environmental assessment (EA) or environmental impact assessment (EIS) will be required.

REFERENCES CITED

Bernero, R.M. 1993 (March 2). Letter from R.M. Bernero, Director, Office of Nuclear Materials Safety and Safeguards, U.S. Nuclear Regulatory Commission, to J. Lytle, Deputy Assistant Secretary for Waste Operations, Office of Waste Management, U.S. Department of Energy, Washington, D.C.

National Research Council. 1995. Technical Bases for Yucca Mountain Standards. Committee on Technical Bases for Yucca Mountain Standards, National Academy Press, Washington, D.C. 205 pp.

National Research Council. 1999. Comments on Proposed Radiation Protection Standards for Yucca Mountain, Nevada, by the Board on Radioactive Waste Management. Board on Radioactive Waste Management, Washington, D.C. 18 pp.

Paperiello, C.J. 1997 (June 9). Classification of Hanford Low-Activity Tank Waste Fraction. Letter from, Director, Office of Nuclear Material Safety and Safeguards, U.S. Nuclear Regulatory Commission, to J. Kinzer, Assistant Manager, Office of Tank Waste Remediation System, Richland Operations Office, U.S. Department of Energy, Washington, D.C.

APPENDIX F

Disposition of the Nevada Test Site

Allen G. Croff

The committee visited a number of U.S. Department of Energy (DOE) sites in the course of this study. The purpose of these visits was twofold: first, to understand better the issues and interrelationships that affect the disposition of various types of DOE sites in differing locations, and second, to acquire information that would permit development by the committee of an integrated approach to site-specific disposition decisions. The Nevada Test Site (NTS) was chosen as one of the sites to be considered by the committee at the request of DOE and because it is representative of a large DOE site where substantial amounts of hazardous materials exist and are likely to remain.

INITIAL STATUS OF THE NTS

The NTS encompasses 3,496 km^2 of land area in southern Nevada reserved for the jurisdiction of the DOE. It features desert and mountainous terrain, and is larger than Rhode Island, making it one of the largest secured areas in the United States. The NTS is in a remote and arid region, mostly surrounded by federal lands, and has strictly controlled access. Some lands are open to public entry. Most of the NTS is located in Nye County, Nevada, with its southernmost point being just 105 km northwest of Las Vegas. The NTS is surrounded by the Nellis Air Force Range (NAFR) Complex on the west, north, and east, and land managed by the U.S. Bureau of Land Management (BLM) on the south and southwest. The NAFR Complex is used for military training. The BLM lands are used for grazing, mining, and recreation. Near the eastern boundary of the NTS, the NAFR Complex shares the use of land with the U.S. Fish and Wildlife Service's Desert National Wildlife Refuge.

The historic activities at the NTS are: atmospheric weapons testing, underground nuclear testing, safety testing of nuclear weapons, nuclear rocket development, near-surface disposal of radioactive wastes, crater disposal of contaminated soils and equipment, greater confinement disposal of radioactive wastes (a term denoting disposal more isolating that near-surface, but not a deep geologic repository (e.g., deep borehole disposal of radioactive wastes such as is used at NTS), and site support activities. From 1951 to 1992 over 820 underground nuclear tests and 100 atmospheric tests were conducted at the NTS (U.S. Department of Energy, 1994). For many of the underground tests, more than one weapon was tested. Ongoing and planned future activities at the NTS include helping to ensure the safety and reliability of the nation's nuclear weapons stockpile, disposal of low-level radioactive wastes, storage of wastes for disposal off site, non-defense research and development (e.g., alternative energy projects, Spill Test Facility, alternative fuels demonstration projects, environmental technology), and use of the site by other Federal agencies for military exercises and R&D projects.

NTS END STATE

DOE/NTS defines "complete clean-up" as bringing a site to the point "that land, facilities, and materials are adequately safe to be available for alternative use, based on future land use policy decisions, with a minimum cost for long-term surveillance and monitoring" (M. Sanchez, 10 September 1997, presentation to the committee). Cleanup priorities and cleanup levels are subject to negotiation with regulators and involved stakeholders.

Stated in broad terms, the presently accepted future use of the NTS is for it to become:

> . . . a diversified national test and demonstration site that can continue to support the reduced nuclear weapons defense program, while also attracting and supporting other high tech programs and industry that can make significant and long term contributions to local and national energy, environmental, defense, and economic needs (Nevada Test Site Economic Adjustment Task Force, 1994).

The DOE and state of Nevada have agreed on a disposition approach that requires residual contamination resulting from nuclear weapon testing be cleaned up to varying degrees consistent with the proposed land use (U.S. Department of Energy, 1996c). Present plans for assessment and remediation of the Nevada Test Site are summarized as follows:

1. Surficial soils that are typically contaminated with uranium and plutonium will be excavated or contained. Currently there is no single national standard establishing cleanup levels for surficial contamination of plutonium; rather, these levels are negotiated on a site-by-site basis between DOE, the state, and the U.S. Environmental Protection Agency (EPA) regional office. However, the final targeted remedial action levels are in the 200 pCi/g range.

2. Underground nuclear weapons testing site remediation plans (IT Corporation, 1998) will involve sequential development of:

- regional and test-specific groundwater flow models for underground test sites that intersect the saturated zone,
- a corrective action investigation plan that could include further characterization of the test sites,
- modeling of test sites for five radionuclides to predict the contaminant boundary and leading to a documented decision concerning the disposition of each test site, and
- a corrective action and closure plan.

3. Other support facilities and the debris from near-surface safety tests will be cleaned up or remediated to a degree yet to be determined based on their potential for future use.

The DOE states that because cost-effective technologies have not yet demonstrated an ability to effectively remove or stabilize radioactive contaminants from the groundwater at the various test sites, subsurface contaminants in and around the underground test cavities will be left as is, and subject to continuing monitoring and surveillance. This approach may be revised if advanced, cost-effective technology is developed. The committee could not find evidence that any such technology development is currently planned by DOE (U.S. Department of Energy, 1997b).

Institutional control of the NTS is assumed in perpetuity at the existing boundaries, and for the foreseeable future, the landlord is assumed to be DOE Defense Programs (U.S. Department of Energy, 1997b, p. 24). However, the defined end state is establishment of a monitoring network, program, and schedule acceptable to DOE, the state, and interested and affected parties, including long-term surveillance and monitoring of the UGTA for a period of 50 years is required (U.S. Department of Energy, 1996c). Extending this period to 100 years is under consideration DOE (U.S. Department of Energy, 1997b, Appendix A).

APPENDIX F

CHARACTERIZATION AND TECHNICAL ASSESSMENT OF THE NTS

Characterization

There are two major aspects of the NTS that require characterization: the contaminant source term and the naturally occurring features of the NTS and surrounding area that are relevant to potential release of or access to the contaminants.

Source Term

Essentially all of the contamination at the NTS results from the radiologically and/or chemically hazardous substances associated with nuclear explosions. The atmospheric radiological source term is composed of volatile species that are released by leakage from historic nuclear weapons test sites and by evaporation. The estimated release rates are 700 Ci/y of tritium and 160 Ci/y of krypton-85 (U.S. Department of Energy, 1996b, Volume 1, p. 4-150).

Surficial radiological contamination is estimated to be 36 Ci. The dominant source of surficial contamination is contaminated soils from nuclear safety tests, but there is also fallout from atmospheric testing. The primary radioelements that were released are plutonium, uranium, and americium, with lesser amounts of cesium, strontium, and europium (U.S. Department of Energy, 1996a, p. 65).

The total underground radiological contamination is about 310 MCi, essentially all from underground nuclear testing. However, the 112 MCi underground radiological source term considered in the NTS environmental impact statement as being available for potential migration is just the total activity from all underground tests that were conducted beneath the water table or within 101 m of the top of the water table, of which about 90 percent is due to tritium (U.S. Department of Energy, 1996a, p. 65). This assumption is apparently based on the belief that the arid nature of the NTS would preclude substantial amounts of radionuclides above this level from mobilizing.[1] In addition, there are substantial uncertainties in the total radiological source term because calculation of the radionuclide composition used estimation, adjustment, and extrapolation techniques to account for (a) significant amounts of radionuclides from testing by Lawrence Livermore National Laboratory, (b) the amount of inventory actually beneath or within 101 m of the water table, and (c) the initial amounts of fissile materials and tritium, and the amount of fission products, actinides, and activation products generated (Borg et al., 1976, p. 79; U.S. Department of Energy, 1996a, p. 74). The Committee has not been able to find any unclassified quantification of these uncertainties, and classified information was not examined in this study.

The toxic materials present after a nuclear weapon detonation occur in three locations: incorporated in the melt glass that pools in the bottom of the cavity, deposited on the rubble and along fractured surfaces within and outside the cavity, and in gases that escape to the atmosphere within a short time after detonation. The distribution of radionuclides is complex, and their behavior during the explosion as well as the chemistry by which they are incorporated or deposited are not fully understood, especially for those species that partition between the melt glass, rubble, and fractures (Borg et al., 1976, p. 177, 187; Kersting, 1996, p. 23; Smith, 1993, pp. 5, 21).

Non-radioactive hazardous materials used in nuclear weapons testing have been surveyed (Bryant and Fabryka-Martin, 1991). Such materials could be introduced into the subsurface from pre- or post-detonation drilling activities, or during sealing of the shot hole before detonation; and as materials used to seal the borehole before detonation. In practice, the non-radioactive hazardous materials typically amount to several tons of lead, a "few kilograms" of other hazardous metals (e.g., arsenic, gallium) and unidentified hazardous organic compounds. It should be noted that nonhazardous organic compounds are also of interest because they may lead to species that complex with hazardous constituents and promote their transport. No unclassified estimates are available concerning the identity and quantity of hazardous and potentially important non-hazardous, non-radioactive materials that

---

[1] DOE officials recently stated that in the future the entire radionuclide inventory would be assumed to be part of the source term (R.M. Bangerter, 1998, personal communication).

may still remain in the subsurface at the NTS. Information regarding the distribution and chemistry of nonradioactive residues that do not have radiological analogues is not evident.

NTS Environment

Atmospheric characterization (e.g., wind direction and frequency, rainfall) related to the transport of gaseous and particulate contamination has been well characterized. However, the mobility of contaminated surficial deposits is less certain. The DOE believes the contaminated soil to be largely gathered around the base of vegetation in immobile positions unless the surface is disturbed (U.S. Department of Energy, 1996a, p. 82), but the basis for this conclusion and its dependence on assumptions concerning future vegetation patterns and surface disturbances are unknown.

In general, the subsurface characteristics (geology, hydrology, geochemistry) of the NTS are not understood at a sufficient level of detail to provide a basis for modeling contaminant transport for the purpose of predicting risks with an acceptable degree of accuracy. This is especially true at Pahute Mesa, which constitutes one of the largest and most difficult hydrogeologic regimes at the NTS (IT Corporation, 1998). This lack of understanding is due, in part, to a combination of the extremely complex and heterogenous geology of the site, and in part to a lack of historical interest in achieving more complete understanding. Recently, attempts to perform two- and three-dimensional hydrologic modeling have been pursued (R.K. Waddell, HIS GeoTrans, Inc., September 10, 1997, presentation to the committee). The data base available to validate these models is meager, but NTS has recently initiated a drilling program for the purposes of subsurface exploration between Pahute Mesa and Oasis Valley, a study of groundwater discharge in Oasis Valley, and a study of water infiltration through test craters (IT Corporation, 1998).

The extent of information and investigation concerning NTS geochemistry is even less than for hydrologic aspects, with the exception of areas having water chemistry and geology similar to that of the Yucca Mountain, which is being extensively investigated as a potential site for a high-level waste repository. While water composition per se is known adequately, the chemistry of its interactions with naturally occurring subsurface materials and characterization of naturally occurring chemicals that might affect radionuclide transport (e.g., colloid formers) is not (Kersting, 1996, p. 25). The DOE has recently initiated geochemical studies between Pahute Mesa and Oasis Valley to determine groundwater age and travel time, and to study colloid transport (IT Corporation, 1998).

Risk Assessment

A risk assessment builds on the foundation provided by the characteristics of the site and source term, and superimposes considerations related to the mobilization, transport, uptake, and impact of contaminants. The important uncertainties and unknowns in these characteristics have been described immediately above, and their implications will not be repeated here. The impact of the other considerations will be discussed below for surficial and subsurface contaminants.

Surficial Contamination

Within the bounds of uncertainty noted above in relation to activities that disturb the soil, the risks from surficial contamination appear to be relatively well understood. The DOE has calculated the maximum effective dose at the site boundary from airborne contaminants to be 0.0048 mrem and the collective effective dose equivalent within 80 km of the NTS to be 0.012 person-rem (U.S. Department of Energy, 1996b, Volume 1, p. 4-152). The risks from various types of habitation of some of the plutonium-contaminated sites are estimated in Daniels (1993, p. 56). Most lifetime risks are low (cancer risk well below 10^{-6}), but for a few sites the risk exceeds 10^{-4}. Within the reports cited, there is no mention of scenarios that involve intrusion or other disturbance of the surface or subsurface.

Subsurface Contamination

There is considerable uncertainty concerning the actual quantity of radioactivity that can be mobilized by leaching of contaminated subsurface debris by groundwater. Smith et al. (1998) have summarized the uncertainties associated with leaching for the NTS and concluded that the radionuclides most likely to become mobile and migrate via the groundwater regime are: (1) tritium; (2) a number of anions and neutral species such as technetium-99, ruthenium-106, chlorine-36, and iodine-129, all assumed to migrate at the same rate as groundwater; and (3) cationic species, including strontium-90, cesium-137, antimony-125, cobalt-60, zirconium-95, plutonium-239, and others, that are believed to move more slowly than groundwater to varying degrees. It should be noted that zirconium-95, ruthenium-106, and antimony-125 all have half-lives less than three years and are not likely to pose a groundwater hazard; the same is probably true for cobalt-60 with a half-life of 5.2 years. However, quantitative estimates are highly uncertain to the point of being almost non-existent. There has been essentially no study of whether the substantial fraction of the radiological source term that was deposited above the water table is moving downward into the saturated zone (Borg, et al., 1976, p. 187; Kersting, 1996, p. 26).

The situation related to retardation of radionuclide transport by sorption onto rocks is somewhat better than for leaching, with several studies having been conducted. Tritium is appropriately assumed to move at the same rate as the groundwater. However, documentation for most other radionuclides indicates that retardation factors vary significantly with respect to water composition, experimental conditions, and rock type. The causes of the variations are speculative (Smith, 1993, p. 18; Kersting, 1996, pp. 23, 25). In fact, a recent study (Daniels, 1993, p. 76) assumed no sorption of any radionuclides because of the limited database.

Otherwise insoluble or highly retarded radionuclides can be transported by forming or attaching to colloidal particles, which then move essentially at the same rate as the groundwater in which they reside. A recent review (Kersting, 1996, p. 24) concluded that a substantial fraction of radionuclides can be associated with colloids, but the effects on transport are not known. Contaminant transport by non-radioactive organic chemicals or degradation products thereof has not been studied or taken into account.

A review of the literature concerning leaching and sorption of radionuclides from nuclear weapons testing melt glasses is given in Smith (1993). The reader should note one important observation from this report: "Most of these investigations were published over ten years ago; the number of tests and access to device debris has diminished during the subsequent decade" (Smith, 1993, p. 24). The committee's investigations support this observation and the continuation of this trend to the present.

Tritium, which is not sorbed and moves at the same rate as groundwater, is the radionuclide considered almost exclusively by DOE in risk analyses. Other radionuclides were assumed by DOE to move very slowly as compared with tritium and, therefore, were not generally considered in the assessments. However, before 1997 about a dozen instances of migration of radionuclides other than tritium have been documented (Nimz and Thompson, 1992). The largest distance of migration of radionuclides other than tritium was not then known to have exceeded 500 m (1,640 ft). Migration of tritium is more difficult to interpret, but is thought to have migrated no more than several kilometers, although tritium, with a half-life of 12.3 years, is not likely to pose a long-term threat to the groundwater resources at NTS.

Pahute Mesa, which is the location of most of the U.S. large nuclear explosions, contains approximately 70 percent of the tritium at the NTS (IT Corporation, 1998). Modeling results also indicate that groundwater flow paths from Pahute Mesa are the shortest of all those at the NTS site and constitute the highest potential for contamination migration to off-site public receptors (IT Corporation, 1998). Recent analysis of water from a well near the TYBO nuclear weapon test site on Pahute Mesa (Thompson, 1998) showed that plutonium as well as cesium, cobalt, and europium were unexpectedly present in the water about 1300 m from the source site associated with the BENHAM test. All of these were shown to be associated with colloidal particles. The plutonium was present at concentrations below drinking water limits (Kersting et al., 1999).

The uptake points for radionuclides are generally assumed to be springs in off-site locations such as Oasis Valley to the southwest of the NTS. This assumption has implications for institutional management of the NTS.

For underground tests conducted within the NTS boundaries, groundwater modeling studies have been performed by Daniels (1993) and GeoTrans (1995). Both of these studies evaluated the migration of tritium from test locations on Pahute Mesa to Oasis Valley. In addition, the GeoTrans study examined migration flow paths from Pahute Mesa to Amargosa Valley and from Yucca Flat to the boundary of the NTS south of Mercury, Nevada. In general, the GeoTrans results for tritium were far below 20,000 pCi/L, which is EPA's allowable tritium concentration in drinking water. The study reported by Daniels (1993) predicted much higher values, some a factor of five less than the drinking water standard. However, these calculations were for screening purposes and used a number of conservative simplifying assumptions. Based on the combined results from these two studies, the estimated range of peak tritium concentrations at the closest uncontrolled use area varies from 5×10^{-4} pCi/L (arriving 150 years after the beginning of migration) to 3,800 pCi/L (arriving in 25 to 94 years). The hypothetical maximally exposed individual at this location is estimated to have a lifetime probability of contracting a fatal cancer between 8×10^{-13} (about one in one trillion) and 1×10^{-5} (about one in 100,000), depending on which model is used.

Very little work has been done on estimating the potential risks from radionuclides other than tritium. Such an attempt was made in Daniels (1993). These estimates are self-characterized as being conservative. The results indicate that at the Area 20 (Pahute Mesa) boundary of the NTS and at Oasis Valley the lifetime committed effective dose for other radionuclides is about 10 percent of that from tritium. Important radionuclides other than tritium were strontium-90, iodine-129, cesium-137, radium-226, plutonium-239, and americium-241. The risks from toxic chemicals used in nuclear weapons tests have not been estimated.

Disposition Alternatives

The DOE has prepared a report (U.S. Department of Energy, 1997a) evaluating the feasibility and cost of selected options for addressing the contamination. An initial list of options was taken from the U.S. Environmental Protection Agency 1994 Remediation Technologies Screening Matrix and Reference Guide (EPA/542/B-94/013). The options in the EPA guide were screened to yield the following list of options:

- No action.
- Intrinsic remediation (reliance on natural subsurface processes).
- Institutional controls.
- Pump and in situ treatment.
- Excavation and on-site treatment and disposal.

All options were determined to be technically feasible, although the "no action" alternative was noted as not meeting EPA requirements for "no action" on a risk basis. All other alternatives were deemed feasible.

Disposition Decision

The NTS is relying on contamination reduction measures for a specific set of contaminated sites such as those having surficial contamination from safety and atmospheric testing and the industrial sites. The goals of most such activities are to reduce contamination levels sufficiently so that the sites do not pose unacceptable risks to inadvertent intruders or during proposed industrial development, but the levels are not sufficiently low to allow site control to cease. The measures generally involve physical removal of contaminated soil and removal of contaminated materials from facilities, followed by burial of the resulting waste. In contrast to this active approach, contamination reduction measures other than natural attenuation for medium-lived species such as tritium are not underway or contemplated for the contamination resulting from underground tests, including the contaminated rock and groundwater.

There is very little reliance by DOE on engineered measures to isolate the contamination at the NTS, especially as it relates to contamination resulting from underground tests. The underground test cavities provide a natural form of isolation that should be well characterized over time regarding migration of radionuclides. These local sites provide information that could be relevant in other arid locations. One exception to this is that DOE has left open the possibility to pump and recycle groundwater if it were to be contaminated with unacceptable levels of tritium at locations accessible to the public. This would presumably continue until radioactive decay made further recycle unnecessary. Other engineering measures for site-wide or high-risk locations have only been studied cursorily (U.S. Department of Energy, 1997a).

To compensate for the relatively small use of contamination reduction and isolation measures, the DOE is placing very heavy reliance on controlling access to the site. The most important of these is to prevent public access to the NTS for the indefinite future, which includes retaining government responsibility for the site and active patrolling to prevent unauthorized site entry. Efforts are also underway to ensure that the activities conducted at "brownfields" within the NTS are consistent with the degree of contamination in particular locations and facilities.

The DOE rationale for assuming indefinite institutional management of the NTS is stated as follows (U.S. Department of Energy, 1997a):

> Institutional controls have been in place at the NTS for over 50 years, and these controls have taken the form of both active and passive; the public knows of the related risks and is aware that the U.S. Government strictly controls access to the NTS. Therefore, because of 50 years of 'Institutional Memory,' it seems reasonable to believe that such controls could continue indefinitely as they complement ongoing clean-up and monitoring efforts.

That active institutional management efforts may prove necessary to maintain such controls is a view reinforced by an earlier report (Daniels 1993, p. 72) that explicitly acknowledges (a) the growth in population in the Las Vegas area and the associated demand for water in an otherwise arid area, and (b) the potential for loss of buffer areas provided by the NAFR lands surrounding much of the NTS that could result from extended cessation of nuclear testing. Both are seen as factors that could increase exposure to hazardous materials presently on or beneath the NTS.

Implementation

Implementation of DOE's currently operative NTS disposition decision is composed of ongoing remedial actions and institutional management measures.

Remediation

The DOE is presently remediating contaminated soils that are near the NTS boundary and have high contaminant concentrations, and also selected facilities for the purpose of reindustrialization. Limited characterization activity (e.g., concerning plutonium migration from the BENHAM test) is underway.

Institutional Controls

The DOE has a comprehensive program for monitoring water and air at locations within and outside the NTS, and the state of Nevada performs independent monitoring. The DOE maintains an extensive guard force to prevent public access to the NTS to prevent exposure to legacy contamination and actively hazardous situations, as well as to protect classified activities.

Future Reconsideration of the Disposition Decision

The committee was unable to identify any specific commitment or process that would result in future re-examination of the major features of site remediation decisions being made today, although decisions will be made on specific details (e.g., cleanup levels for specific locations) on a continuing basis. There appears to be little driving force for such reconsideration at present. Thus, the destiny of the site appears to be a limited number of remedial actions consistent with re-industrialization in selected portions of the site, followed by an indefinite period of institutional control.

REFERENCES CITED

Borg, I.Y., R. Stone, H.B. Levy, and L.D. Ramspott. 1976 (May 25). Information Pertinent to the Migration of Radionuclides in Ground Water at the Nevada Test Site—Part I: Review and Analysis of Existing Information. Lawrence Livermore Laboratory UCRL-52078, Lawrence, Calif. 216 pp.

Bryant, E.A., and J. Fabryka-Martin. 1991 (February). Survey of Hazardous Materials Used in Nuclear Testing. Los Alamos National Laboratory LA-12014-MS, Los Alamos, N.M. 12 pp.

Daniels, J.I. (ed). 1993 (June). Pilot Study Risk Assessment for Selected Problems at the Nevada Test Site (NTS). Lawrence Livermore National Laboratory University of California UCRL-LR-113891, Livermore, Calif. 97 pp.

GeoTrans, Inc. 1995 (August). A Fracture/Porous Media Model of Tritium Transport in the Underground Weapons Testing Areas, Nevada Test Site. GeoTrans, Inc., Boulder, Colo. 62 pp.

IT Corporation. 1998 (March). Corrective Action Unit Modeling Approach for the Underground Test Area, Nevada Test Site, Nye County, Nevada. Report DOE/NV/13052-501, Las Vegas, Nev.

Kersting, A.B. 1996 (December). The State of the Hydrologic Source Term. Lawrence Livermore National Laboratory UCRL-ID-126557, Lawrence, Calif. 30 pp.

Kersting, A.B., D.W. Efurd, D.L. Finnegan, D.J. Rokop, D.K. Smith, and J.L. Thompson.1999 (January 7). Migration of plutonium in ground water at the Nevada Test Site. Nature 397:56-59.

Nevada Test Site Economic Adjustment Task Force. 1994 (June). State of Nevada Plan of Action for the Future of the Nevada Test Site and Its Work Force. State of Nevada Commission on Economic Development, Las Vegas, Nev. 24 pp.

Nimz, G.J., and J.L.Thompson. 1992. Underground Radionuclide Migration at the Nevada Test Site. U.S. Department of Energy Nevada Field Office DOE/NV-346, UC-703, Las Vegas, Nev. 17 pp.

Smith, D.K. 1993 (May). A Review of Literature Pertaining to the Leaching and Sorption of Radionuclides Associated with Nuclear Explosive Melt Glasses. Lawrence Livermore Laboratory UCRL-ID-113370, Lawrence, Calif. 26 pp.

Smith, D.K., A.B. Kersting, T.P. Rose, J.M. Kenneally, G.B. Hudson, G.F. Eaton, and M.L. Davisson. 1998 (May). Hydrologic Resources Management Program and Underground Test Operation Unit FY 1997 Progress Report. Lawrence Livermore National Laboratory UCRL-ID-130792, Livermore, Calif. 92 pp.

Thompson, J.L. (ed). 1998 (February). Laboratory and Field Studies Related to the Radionuclide Migration at the Nevada Test Site: October 1, 1996-September 30, 1997. Los Alamos National Laboratory LA-13419-PR Progress Report, Los Alamos, N.M. 31 pp.

U.S. Department of Energy. 1994 (December). United States Nuclear Tests, July 1945 through September 1992. Nevada Operations Office, Office of External Affairs DOE/NV-209 (Rev. 14). Las Vegas, Nev.

U.S. Department of Energy. 1996a (August). Geology, Soils, Water Resources, Radionuclide Inventory; Technical Resources Report for the Final Environmental Impact Statement for the Nevada Test Site and Off-Site Locations in the State of Nevada. Nevada Operations Office, Las Vegas, Nev. 123 pp.

U.S. Department of Energy. 1966b (August). Final Environmental Impact Statement for the Nevada Test Site and Off-Site Locations in the State of Nevada. Nevada Operations Office DOE/EIS-0243.

U.S. Department of Energy. 1996c (December 9). Record of Decision: Environmental Impact Statement for the Nevada Test Site and Off-Site Locations in the State of Nevada. Secretary, Department of Energy, 6450-01-P, Washington, D.C. 56 pp.

U.S. Department of Energy. 1997a (April). Focused Evaluation of Selected Remedial Alternatives for the Underground Test Area. Nevada Operations Office DOE/NV-465, Las Vegas, Nev.

U.S. Department of Energy. 1997b (June). Nevada Operations Office Environmental Management Accelerating Cleanup: Focus on 2006—Site Discussion Draft. Nevada Operations Office, Las Vegas, Nev. 31 pp.

APPENDIX G

Mathematical Models Used for Site Closure Decisions

Shlomo P. Neuman and Benjamin Ross

The U.S. Department of Energy (DOE) faces difficult decisions concerning the disposition or closure of sites contaminated with radioactive, toxic, and hazardous materials. Given current knowledge and technology, it is neither economically nor technically feasible to release all DOE sites for unrestricted use in the foreseeable future. It will therefore be necessary to keep many sites under some form of control well into the future. The DOE is considering long-term stewardship to encompass all activities that are required to maintain an adequate level of protection to human health and the environment from hazards posed by nuclear and chemical materials, waste, and residual contamination remaining after cleanup is completed. As part of its decision process the DOE will need to assess the consequences of alternative remediation, restoration, control, and/or release scenarios at each site. In particular, it will need to assess potential risks and hazards posed to human health and the ecology by contaminants that remain at a site following remediation and restoration, regardless of whether the site is released or remains under DOE control. This includes assessing the long-term performance of engineered barriers to contaminant migration at the site.

Assessments of the hazards posed by sites where contamination will remain into the distant future are known as **risk assessments** or **performance assessments**. The term performance assessment usually refers to evaluation of the extent to which an engineered system satisfies predetermined design or performance criteria. In the context of contaminated sites, the system of concern usually includes both engineered and natural components, and performance criteria relate both to the design of engineered remedies and to human and ecological safety measures. Such safety measures may (but need not) be cast in the form of risk criteria; in the latter case, one speaks of risk assessment. Any risk or performance assessment uses mathematical models, usually but not always implemented on computers, which describe the processes that operate at the site. The models rely, however, on information about the site, including its physical properties and the pathways of human exposure to contamination. This information determines what parts of the mathematical models are deemed relevant and what parameter values and forcing terms (e.g., source terms, initial and boundary conditions) are input as data.

PRESENTLY AVAILABLE MODELS

Two kinds of models are typically used to predict the behavior of a site contaminated with radioactivity or toxic chemicals:

1. a hydrologic transport model that predicts how dissolved contamination will be transported in groundwater; and

2. a "risk" model that computes the transfer of contaminants through different portions of the surface environment, the exposure of humans to contaminants in the environment, and the resulting health effects.

These models can be supplemented with a variety of other models, such as:

- for radioactive contamination, a direct exposure model that computes the dose to humans from radiation emitted by contamination in the ground (this pathway does not exist for chemical contaminants);
- a leaching model that describes how contamination passes from the solid phase into the aqueous phase;
- a vadose zone model that describes how contamination moves downward from the point of disposal toward the water table, or upward with some contaminants such as radon;
- an air dispersion model that describes the transport of dust that blows off contaminated soil, or of gaseous contaminants such as radon; and
- an ecological risk model that evaluates how the contamination affects ecosystems.

Most commonly, models used at DOE remediation sites involve direct exposure, hydrologic transport, and risk. Modeling can be done with separate models for each part of the problem, or one model can handle the entire problem. Prominent among the multimedia or multiple-pathway risk assessment models that try to carry out all steps in risk assessment in a single model are RESRAD (Yu et al., 1993; Chen et al., 1991; Chen et al., 1995), MMSOILS (U.S. Environmental Protection Agency, 1996; Chen et al., 1995), and Multimedia Environmental Pollutant Assessment Systems (MEPAS) (Buck et al., 1995; Buck et al., 1997; Chen et al., 1995; Doctor, et al., 1990; Streile et al., 1996; Whelan, et al., 1996).

Direct Exposure to Radiation

Waste units where direct exposure to radiation is the major hazard are frequently modeled with the RESRAD code. This model was initially developed to implement DOE's Residual Radioactive Material Guidelines and the U.S. Nuclear Regulatory Commission (USNRC) procedures to assess site decommissioning. It was subsequently expanded by incorporating a risk model and simple models of hydrologic transport and leaching. RESRAD is used heavily in DOE decision-making. For example, it features prominently in a recent DOE document (U.S. Department of Energy, 1996), which addresses remedial designs and remedial actions for high-priority waste sites in the 100 Area of Hanford. The same document is expected to form the base for remedial actions across the 100 Area liquid waste disposal sites with an intention to revise it for future remedial actions. RESRAD has also been used extensively in decision-making about cleanup of areas in Nevada that were contaminated with plutonium by testing of nuclear weapons. Because it comprehensively implements the DOE and USNRC guidelines and has been thoroughly tested, RESRAD is a reliable tool for solving direct exposure problems. DOE's reliance on RESRAD at sites where the major hazard is ground shine or inhalation of radioactive dust (also addressed by the DOE and USNRC guidelines that RESRAD implements) is quite appropriate. However, as discussed below, the other subunits of RESRAD cannot be relied upon in the same way.

"Risk" Models

So-called "risk" models actually carry out only a part of the computations that go into a risk assessment. These models identify pathways of exposure and calculate human intake, dose, and detriment. They generally take the concentrations of contamination in soil and surface waters as an input; these quantities must be measured, calculated by a separate model, or in the case of an integrated performance assessment model calculated by a separate submodel.

Essentially, the risk models implement the Risk Assessment Guidance for Superfund, which combines a linear "box model" of ecosystem transfers with coefficients published by the U.S. Environmental Protection Agency

(EPA) that give the harm per unit of chemical contaminant ingested by a human being. The coefficients for carcinogens are based on a linear no-threshold model of detriment; for non-carcinogens, there is assumed to be a threshold below which no harm occurs. For radionuclides, risk coefficients are derived from human exposure data and are published by the International Commission on Radiological Protection (ICRP) and the National Council on Radiation Protection and Measurements (NCRP).

Ecosystem transfers are usually modeled as a linear system. "Default" values for the transfer coefficients that define this linear system have also been published by EPA and USNRC. There are so many of these transfer coefficients that it is impossible to measure them all, so use of the default values is essential, but these values will not always be correct. The proper practice is to adopt the default values for pathways of little importance, but to take care to base transfer coefficients for the dominant exposure pathways on site-specific information.

Risk modeling can be carried out either with a computer program or on a spreadsheet. In non-DOE contamination sites, spreadsheet analysis is more common. A recent study commissioned by DOE (Regens et al., 1999) has evaluated RESRAD, MMSOILS and MEPAS and compared them with the spreadsheet approach. It found that the computer models had little or no practical advantage over spreadsheets in usability and efficiency.

Groundwater Transport Models

It is now widely recognized that the subsurface is a complex, multiscale, spatially variable natural environment that can never be fully characterized. Hence the results of even the most thorough site characterization and monitoring efforts are ambiguous and uncertain. To address uncertainties, it has become common to analyze hydrogeologic data statistically and flow and transport stochastically. The most common and straightforward method of stochastic flow and transport analysis involves repeated simulations by means of detailed numerical models in which the material properties (such as permeability and porosity) and forcing terms (sources and boundary conditions) vary randomly from one simulation to another. Permeability and porosity are known to be spatially auto- and cross-correlated on a variety of scales. By taking account of such correlations, and forcing the random variables to conform to measurements, one obtains conditional Monte Carlo solutions to the stochastic flow and transport problems. Upon averaging these solutions one obtains optimum unbiased predictors of system behavior under uncertainty. Upon calculating the variance of the Monte Carlo solutions one obtains a measure of predictive uncertainty.

Vadose Zone Models

Virtually all existing multimedia risk assessment models view fluid flow and radionuclide transport in the vadose zone as moving vertically downward at a uniform and steady rate. Though many recognize that this conceptual model is oversimplified, it is often defended as being conservative, in that mathematical models based upon it overpredict contaminant concentrations at receptor locations and associated risks. Reliance on vadose zone monitoring is important in arid and semiarid environments where unsaturated soil conditions may prevail to considerable depths, as at Hanford Site in Washington, the Idaho National Engineering and Environmental Laboratory, and the Nevada Test Site (including Yucca Mountain, the site currently being evaluated as a potential geological repository for high-level wastes and commercial spent nuclear fuel). It is much less important in moderate and humid environments where the vadose zone tends to be shallow and hydrologic variables can be monitored effectively, with relative ease, at and below the water table.

Leaching Models

The transfer of contaminants from the immobile soil phase to groundwater is generally modeled with relatively simple analytical expressions. The choice among these expressions depends on the physical and chemical form of the contamination. Radioactive wastes are generally solids. Two commonly used models are the "leach-limited" and "solubility-limited" models. In the "leach-limited" model, the radionuclides are considered to be incorporated into a solid matrix (crystalline or non-crystalline) that releases minor impurities into groundwater as it alters or dissolves. All radioactive species in the matrix are released in proportionate amounts. In the "solubility-limited" model, the concentration of each radioelement in groundwater is equal to or less than its solubility.

Similarly there are two alternative models for dissolution of organics. If a contaminant is adsorbed to soil particles, the concentration in groundwater will be proportional to the concentration in the soil. On the other hand, if a separate non-aqueous phase liquid (NAPL) is present, the concentration in groundwater in direct contact with the NAPL will be equal to the compound's "effective solubility." The effective solubility is, approximately, the product of the solubility of the pure compound in water multiplied by the fraction of the NAPL that the compound constitutes. When, as is usual, the NAPL is present in disconnected zones of residual contamination, there will be dilution due to the fact that only some of the water that passes through the source will come into direct contact with the NAPL.

CONSTRAINTS AND LIMITATIONS OF MODELS

The models used in support of site decisions are necessarily imperfect reflections of the real environment. Some major limitations of currently available models are described in this section.

Risk Assessment Models

The 1998 Consortium for Environmental Risk Evaluation (CERE) study (Regens et al., 1999) examined the applicability of multimedia risk assessment models to real DOE sites through two case studies using actual data. One site examined was a solid waste storage area at Oak Ridge National Laboratory (ORNL) in Tennessee with trenches containing alpha-contaminated low-level waste, remote-handled transuranic wastes deposited in concrete casks and combination (wood/metal) boxes, and a small number of steel drums. The other case study concerned Operable Unit 2 at the Rocky Flats Environmental Technology Site (RFETS) in Colorado, which contains drums of radioactive-contaminated oils and solvents, plutonium-239 contaminated soils, liquid chemical waste in disposal trenches, and an inactive Reactive Metal Destruction site. The model comparison indicated that 1) the exposure and risk assessment frameworks in all three models follow DOE, EPA and USNRC guidelines; 2) existing major differences between the models are due to their differing objectives—where the capabilities of the models overlap, such differences are due to the formulation of transport components; 3) the models yield results that differ by up to three or four orders of magnitude; 4) the models are in many ways similar to traditional spreadsheet analyses; and 5) the primary benefit of the screening-level risk assessment process is to identify chemicals and pathways that make the largest contributions to overall risk.

Spreadsheets (or paper-and-pencil calculations) are much more flexible than computer models. For example, all existing multimedia models consider a single source for each surface water pathway. At large DOE sites like ORNL and the Savannah River Site (SRS) in South Carolina, contaminant loading to surface water bodies is likely to involve creeks and rivers intersecting several contaminant plumes at various locations, and surface runoff from multiple sources may impact a single stream at several points. Other calculations that existing models cannot handle include the combination of stream flows and contaminant loadings as tributary creeks flow into larger creeks and streams, sediment uptake, and contaminant decay processes.

Another great disadvantage of the computer models is that the assumptions (where things most commonly go wrong) are buried in the computer code. This creates a strong presumption in favor of default assumptions, which can easily go wrong. For example, if local populations engage in subsistence fishing, the default value will underestimate fish consumption and lead the modeler to overlook an important pathway due to bioaccumulation in fish. If vacation homes have water intake pipes that lie on the bed of a lake, default assumptions about mixing in the lake will cause the homeowners' exposure to groundwater that discharges into the lake to be underestimated. The way to uncover such mistakes is to have the widest possible review and criticism. Review by local community members, who are often more familiar with the realities of a site than outside experts, is especially valuable. The effect of using a computer program rather than a spreadsheet (or paper-and-pencil calculation) to do the risk assessment is that the assumptions that most need review are hidden where they are least accessible.

In general, CERE found that the advantages of multimedia models are not as great as anticipated. Differing objectives and lack of transparency make model application difficult; application to real situations may require considerable ingenuity and expertise on the part of the user. CERE also observed that multimedia models do not provide absolute estimates of risk, but rather conditional estimates based on multiple assumptions about source term, environmental settings, transport characteristics, exposure scenarios, toxicity, and other variables. While the magnitude of risk estimates produced by multimedia models differ, they do tend to agree on the most significant contaminants and the most important pathways of exposure. These observations would be equally applicable to spreadsheets and other methods of risk assessment that do not rely on computer programs.

Groundwater Transport Models

In principle, the Monte Carlo method of uncertainty analysis should be easy to implement in conjunction with a risk assessment methodology of the kind just described. The only potential obstacle for such implementation is the large amount of computer time that may be required to repeat detailed hydrologic model simulations many times so as to generate a meaningful statistical sample of equally likely random flow and transport solutions. The large amount of computer time required by conditional Monte Carlo simulations conducted by means of detailed, state-of-the-art hydrologic models is often cited as a justification for either foregoing such simulations completely (and with them, the opportunity to quantify prediction uncertainty) or for using highly simplified models. It is the consensus of many hydrologists that, given the critical importance of groundwater flow and transport models in assessing risks and hazards from subsurface contamination, it is better to run a small number of simulations with detailed models that incorporate the known physics and geology of the sites rather than a large number of simulations with oversimplified models that may disregard crucial information.

Deterministic models are unable to account for uncertainties in input data and therefore yield outputs (such as contaminant concentrations, exposure doses and risks) of unknown reliability. Without providing quantitative information about the uncertainty (hence reliability) of its outputs, a model cannot be used to assess 1) the worth of additional data through site characterization, 2) the reliability of a proposed environmental monitoring system, or 3) the uncertainty associated with predicted site performance measures (such as future contaminant concentrations, doses, and risks). Hence, uncertainty analysis must be an integral part of future performance or risk assessment effort by the DOE. When the main uncertainties are quantifiable, the simplest way to accomplish this is to operate the corresponding models in a conditional Monte Carlo mode as described earlier.

But some major sources of uncertainty are difficult to quantify. A model may reflect an inaccurate conceptual-mathematical representation of site hydrology and subsurface transport processes. For example, long-established conceptual and mathematical models of groundwater flow have come into question at the Nevada Test Site. Groundwater in southern Nevada flows long distances, often passing through several topographic basins between recharge and discharge. For many years, models of this system were based on a conceptual framework originally established by Winograd and Thordarson (1975), who were unable to determine the northern boundary of the flow system because they had very few data north of the test site. Maps in their report ended at 38°N latitude, well beyond the limit of their data. Subsequent studies that used isotopic variations to infer the origin of groundwater considered only recharge areas within the boundaries of the Winograd and Thordarson maps. Water found in parts of the test site with low concentrations of oxygen-18 and carbon-14 was interpreted as water that had recharged in the pluvial conditions of the late Pleistocene, about 10,000 years ago (Claassen, 1985). This conclusion implied that groundwater moves very slowly in the test site area. Recently, Davisson et al. (1999) proposed a new interpretation in which most water with low oxygen-18 concentrations originated in recharge areas north of the area studied by Winograd and Thordarson. The low carbon-14 content of this water is explained in this view by isotopic exchange with carbonate rocks. This interpretation suggests much greater speeds of groundwater movement. Whichever interpretation of the southern Nevada flow system ultimately turns out to be correct, this story illustrates how a concept initially introduced as an unverified simplification can become embedded in scientific thinking as an unexamined assumption that greatly influences conclusions.

As another example, actinides such as plutonium and americium are strongly adsorbed or insoluble in laboratory experiments, and most computer models assume that actinides only move when they are in a dissolved state. The assumption of thermodynamic equilibrium between dissolved and adsorbed phases implies that actinides move very slowly in the subsurface. But sorption on colloidal particles of clay, silica, or organic material may significantly enhance their mobility. Two wells completed in the vicinity of the TYBO underground nuclear test site on Pahute Mesa, at the Nevada Test Site, were sampled as they were pumped. The sampling revealed the presence of plutonium, in association with colloids, at significant concentrations in well ER-20-5 #1, 278 m west of TYBO at a depth of 860 m, and at very small concentrations in a deeper aquifer penetrated by well ER-20-5 #3, 30 m south of #1 at a depth of 1,309 m (Kersting et al., 1999).

Vadose Zone Models

To better understand fluid flow and contaminant transport processes in the vadose zone, one must recognize that unsaturated soils and rocks form part of a complex three-dimensional, multiphase, heterogeneous, and anisotropic hydrogeologic system. This system does not constitute a perfect sequence of horizontal layers with homogeneous properties as would be needed for flow and transport to be uniform in the vertical direction. If it did, flow and transport rates would be controlled by the least permeable layer and would therefore be correspondingly low.

In reality, unsaturated medium properties vary spatially in a complex manner, which often allows fluids and contaminants to move around low-permeability obstacles much faster than would be possible in the perfectly stratified case. Preferential flow through high-permeability channels, the formation of unstable fingers, and development of fractures can further enhance the rate of contaminant migration from a source in the vadose zone to the water table. Preferential flow and fingering have been widely documented in laboratory and field studies, demonstrated numerically, and explained theoretically (Chen et al., 1995). Birkholzer and Tsang (1997) have shown numerically that solutes in randomly heterogeneous unsaturated soils migrate rapidly along narrow channels, which are random and vary dynamically with the flow regime. Ignoring these and other phenomena such as the intermittence of infiltration, by assuming that flow is perfectly uniform and vertical as is done in existing multimedia models, renders these models nonconservative in that they underestimate (rather than overestimate, as claimed erroneously by their adherents) contaminant mass flow rates through the vadose zone. Another complicating factor that needs to be considered at more humid sites such as ORNL and SRS is the possibility that contaminants could seep laterally through the soils in a shallow unsaturated zone and into small surface depressions as has been observed in the field, and explained theoretically, by Zaslavsky and Sinai (1981). In addition, flow and contaminant transport in the vadose zone are not always directed downward toward the water table.

A panel of four experts concluded that characterization of the vadose zone is an essential step toward understanding contamination of the groundwater, assessing the resulting health risks, and defining the concomitant groundwater monitoring program needed to verify the risk assessments (Conaway et al., 1997). The panel concluded that reliable computer models of groundwater contamination could not be developed without reliable data on the transport of contaminants within the vadose zone. As that subject is poorly understood, previous and ongoing computer modeling efforts are inadequate and based on unrealistic and sometimes optimistic assumptions that render their output entirely unreliable.

Downward migration from the Hanford Site tanks provides a strong warning about the dangers of oversimplifying the vadose zone (U.S. Department of Energy, 1998). Because DOE had assumed that wastes would move slowly, if at all, through the vadose zone, it never issued a comprehensive plan to assess vadose zone conditions at Hanford and funded few studies of flow or transport through it. Experts have repeatedly advised DOE that its concept of vadose zone hydrology had been potentially flawed, but the expert advice remained unheeded for a long time.

Beginning in 1994, DOE's Grand Junction Office, using technology developed to detect uranium ore deposits, performed tests in about 800 boreholes in the single-shell farm at Hanford. The tests were intended to provide baseline information about the distribution of certain radioactive wastes, but they also enabled the team to identify radioactive substances at considerable depths in the vadose zone. The team found indications of possible new leaks in some tank farms and deep contamination by some radionuclides in several farms. Cesium was discovered at a depth of 125 ft below one single-shell tank farm, and just above the water table under another tank farm. After deepening the well near the first farm, DOE found cesium at a depth of 142 ft and technetium at a depth of 177 ft (Rust Geotech, 1996). A study by the Los Alamos National Laboratory has shown leaks at one farm to be three to six times greater than previously reported. A January 1988 report by Pacific Northwest National Laboratory has shown that wastes from one farm have reached groundwater.

In December 1997, the DOE announced publicly that highly radioactive wastes from previously leaking underground storage tanks had migrated all the way down to groundwater. DOE now acknowledges that there are significant uncertainties and data gaps in its understanding of the inventory, distribution, and movement of contaminants in the vadose zone at Hanford. Yet the agency is only now starting to develop a comprehensive strategy for investigating the vadose zone (U.S. Department of Energy, 1998, 1999).

Leaching Models

To properly model leaching of a contaminant into groundwater, one must select a model that corresponds to the physical and chemical state of the contaminant. Order-of-magnitude errors can result if this is done incorrectly. The dissolution of radioactive wastes will be underestimated if the mineral that is assumed to limit solubility does not precipitate, either because of kinetic constraints or because the oxidation state of the element has not been correctly identified. Organic contaminant dissolution is frequently modeled by an adsorption-based equation. When NAPL is present, the adsorption-based equations may greatly overestimate the dissolution rate. Johnson et al. (1990) observe that the NAPL model is almost always better for hydrocarbon spills, and comment that the frequent use of the adsorption equation in modeling is "due to its mathematical characteristics, rather any model validation. . . ."

DISCUSSION

Compatibility of Models with Measurements

It is essential to ensure that models are consistent with field measurements of environmental variables. This is particularly important in using multimedia models, whose input and output variables frequently are not directly observable. In order to assess potential risks and hazards from residual contaminants under various cleanup and land/water use scenarios, one should ideally have detailed information about their nature, quantity and location; the manner and rate at which they could be mobilized to migrate toward human and/or ecological receptors; the pathways and rates of their migration; their concentrations at receptor locations; associated doses to receptors; and their effects on receptor health. In reality, information about current site conditions is limited and so is the ability of models to predict future conditions at most sites. These limitations stem from the fact that environmental and bioecological processes, which control contaminant behavior and its health effects at most sites, are extremely complex and therefore exceedingly difficult to describe.

The simplified multimedia models described above often have hidden built-in assumptions that will lead to errors at sites where they do not apply. Because such a model can neither be applied directly to real data nor confirmed experimentally, it is difficult to apply correctly and nearly impossible to evaluate. Use of multimedia models should be confined to problems where the multimedia models incorporate a state-of-the-art submodel (such as direct exposure to gamma radiation in RESRAD) or where assumptions about non-measurable variables are imposed by regulatory fiat (such as the cancer risk factors determined by EPA).

The principle of parsimony should be used to differentiate between alternative operational models. This principle states that among all operational models that one can use to explain a given set of experimental data, one should select the model that is conceptually least complex and involves the smallest number of unknown (fitting) parameters. (This principle can also be stated under a scientific and philosophic rule known as Occam's razor, stating that the simplest of compelling theories should be preferred to the more complex.) When the database is limited and/or of poor quality, one has little justification for selecting an elaborate model with numerous parameters. Instead, a simpler model should be preferred that has fewer parameters, which nevertheless reflects adequately the underlying hydrogeologic structure of the system, and the corresponding flow and transport regime. An inadequate model structure (conceptualization) is far more detrimental to its predictive ability than is a suboptimal set of model parameters.

Risk, Values, and Decision-Making

Decisions that balance risk against cost and other values are among the hardest choices that public officials are called on to make. The difficulties of measuring and communicating risk compound the difficulties created by the need to balance incommensurate values held by different individuals and even within the same individual.

When day-to-day decisions are made about known present-day exposures to chemicals or radiation, the difficulties of doing a risk assessment are frequently avoided by relying on exposure guidelines. The difficult balancing of risk, cost, and uncertainty has already been done by the regulatory agency that set the guidelines. However, decisions about site disposition involve future risks, where not only is the effect on human health of an exposure uncertain, but it is impossible to know whether the exposure will even occur. Thus, while regulatory guidelines can be very useful for making decisions, especially where a conservative analysis predicts exposures below present-day limits, they cannot solve all problems.

The complexities of risk assessment have been the theme of a series of National Research Council reports, including *Risk Assessment in the Federal Government: Managing the Process* (National Research Council, 1983) and *Science and Judgment in Risk Assessment* (National Research Council, 1994). A recent report entitled *Understanding Risk: Informing Decisions in a Democratic Society* (National Research Council, 1996) directly addressed the question of how risk assessments can be made useful in public decision-making. This report concludes that "risk characterization should be a decision-driven activity, directed toward informing choices and solving problems." The report emphasizes the need for risk characterization to consider the values and interests of all interested and affected parties. It describes risk characterization not as a purely technical analysis, but as:

> the outcome of an analytic-deliberative process. Its success depends critically on systematic analysis that is appropriate to the problem, responds to the needs of interested and affected parties, and treats uncertainties of importance to the decision problem in a comprehensible way. Success also depends on deliberations that formulate the decision problem, guide analysis to improve decision participants' understanding, seek the meaning of analytic findings and uncertainties, and improve the ability of interested and affected parties to participate effectively in the risk decision process. The process must have an appropriately diverse participation or representation of the spectrum of interested and affected parties, of decision-makers, and of specialists in risk analysis at each step.

The imperfections of risk assessment as a tool for predicting the long-term behavior of wastes in the sites makes this recommendation particularly relevant for decision-making about site disposition. Because calculations of long-term risk necessarily rely on unverifiable assumptions about the future behavior of people and institutions, it is essential that the assumptions made in the analysis are widely understood by and acceptable to the parties involved in the decision.

Closing Remarks

There has been a tendency by the DOE and some other agencies to rely excessively on models in the context of waste disposal and site contamination issues. Models have been used repeatedly to "demonstrate" that a potential waste disposal site or remedial option complies with regulations and is therefore "safe." More often than not, the ability of models to provide such safety assurances has been taken for granted without a serious attempt to validate them against site data. This is especially true about one-dimensional "multimedia" or "multiple-pathway" dose and/or risk assessment models such as RESRAD, MMSOILS, MEPAS, and DandD (Beyeler et al., 1998; Gallegos et al., 1998), which are based on a limited menu of highly simplified conceptual models, are often used (for screening as well as more advanced investigative purposes) with generic parameters and inputs rather than with site-specific data, and are virtually never compared against actual site conditions. It is however also true, albeit to a lesser extent, about more complex two- and three-dimensional subsurface flow and contaminant transport models that incorporate various details of site geology. The tendency has been to rely on models at the expense of detailed site investigations, site monitoring, and field experimentation. In fact, models have often been used to "demonstrate" that additional site or experimental data would be of little value for a project. The reasons for this state of affairs are easily identified as regulatory and budgetary pressures.

It is often tempting to "demonstrate" by means of a model that a given waste disposal or remedial option is safe, or that additional site data would be of little value, by basing the model on assumptions, parameters and inputs that favor a predetermined outcome. A common example of such bias is the assignment of lower permeabilities to a groundwater flow model than is justified by available data. It is likewise tempting to help a model appear credible by basing it on a unique system conceptualization and subjecting it to sensitivity and uncertainty analyses in which parameters and input variables are constrained to vary within narrower ranges of values than is warranted by the available information. Such practices are common and ultimately detract from the credibility of agencies that employ them.

CONCLUSIONS

- Models are appropriate, often essential, tools for risk assessment and decision-making concerning cleanup and management of contaminated, or potentially contaminated, sites. However, it is inappropriate to use models as "black boxes" without tailoring them to site conditions and basing them firmly on site data. Neither disregard of models nor overreliance on them are desirable.

- The environment constitutes a complex system that can be described neither with perfect accuracy nor with complete certainty. It is imperative that uncertainties in system conceptualization and model parameters and inputs be properly assessed and translated into corresponding uncertainties in risk and decisions concerning risk management. The quantification of uncertainties requires a statistically meaningful amount of quality site data. Where sufficient site data are not obtainable, uncertainty must be assessed through a rigorous critical review and sensitivity analyses.

- Models and their applications must be transparent to avoid hidden assumptions. Model results must not be accepted blindly because hidden assumptions are easily manipulated to achieve desired outcomes.

- Decisions concerning site disposition and risk management should account explicitly and realistically for lack of information and uncertainty.

- The monitoring of site conditions and contamination is an imperfect art. It is important that uncertainty associated with monitoring results be assessed a priori and factored explicitly into site remedial design and post-closure management.

- Where effective and affordable science and technology are not readily available for site characterization, remediation, monitoring, and analyses, the DOE should initiate and pursue vigorously a suitable research and development program. The goals of this program should be both short- and long-term. The program should engage a broad array of talents and specialties from government, industry, and academia in order to maintain a proper balance between disciplines and basic as well as applied research.

REFERENCES CITED

Beyeler, W.E., T.J. Brown, W.A. Hareland, S. Conrad, N. Olague, D. Brosseau, E. Kalimina, D.P. Gallegos, and P.A. Davis. 1998 (January 30). Review of Parameter Data for the NUREG/CR-5512 Residual Farmer Scenario and Probability Distributions for the DandD Parameter Analysis. Letter Report for NRC Project JCN W6227, U.S. Nuclear Regulatory Commission, Washington, D.C.

Birkholzer, J., and C-F. Tsang. 1997 (October 1). Solute channeling in unsaturated heterogeneous porous media. Water Resources Research 33(10):2221-2238.

Buck, J.W., G. Whelan, J.G. Droppo, Jr., D.L. Strenge, K.L. Castleton, J.P. McDonald, C. Sato, and G.P. Streile. 1995. Multimedia Environmental Pollutant Assessment System (MEPAS) Application Guidance. Pacific Northwest National Laboratory PNL-10395, Richland, Wash.

Buck, J.W., D.L. Strenge, B.L. Hoopes, J.P. McDonald, K.J. Castleton, M.A. Pelton, and G.M. Gelston. 1997. Description of Multimedia Environmental Pollutant Assessment System (MEPAS) Version 3.2 Modification for the Nuclear Regulatory Commission. U.S. Nuclear Regulatory Commission NUREG/CR-6566, Washington, D.C.

Chen, J.-J., C. Yu, and A.J. Zielen. 1991. RESRAD Parameter Sensitivity Analysis. Argonne National Laboratory ANL/EAIS-3, Argonne, Ill.

Chen, J.-J., J.G. Droppo, E.R. Failace, E.K. Gnanapragasam, R. Johns, G. Laniak, C. Lew, W. Mills, L. Owens, D.L. Strenge, J.F. Sutherland, C.C. Travis, G. Whelan, and C. Yu. 1995. Benchmarking Analysis of Three Multimedia Models; RESRAD, MMSOILS, and MEPAS. U.S. Department of Energy DOE/ORO-2033, Washington, D.C.

Chen, G., M.Taniguchi, and S.P. Neuman. 1995 (May). An Overview of Instability and Fingering During Immiscible Fluid Flow in Porous and Fractured Media. Report NUREG/CR-6308, prepared for U. S. Nuclear Regulatory Commission, Washington, D.C.

Claassen, H.C. 1985. Sources and Mechanisms of Recharge for Ground Water in the West-Central Amargosa Desert, Nevada: A Geochemical Interpretation. U. S. Geological Survey Professional Paper 712-F, Washington, D.C. 31pp.

Conaway, J.G., R.J. Luxmoore, J.M. Matuszek, and R.O. Patt. 1997 (April). Tank Waste Remediation System Vadose Zone Contamination Issue: Independent Expert Panel Status Report. DOE/RL-97-49 Rev. 0, Richland, Wash.

Davisson, M.L., D.K. Smith, and T.P. Rose. 1999. Isotope hydrology of southern Nevada groundwater: Stable isotopes and radiocarbon. Water Resources Research 35(1):279.

Doctor, P.G., T.B. Miley, and C.E. Cowan. 1990. Multimedia Environmental Pollutant Assessment System (MEPAS) Sensitivity Analysis of Computer Codes. Pacific Northwest Laboratory PNL-7296, Richland, Wash.

Gallegos, D.P., T.J. Brown, P.A. Davis, and C. Daily. 1998. Use of DandD for Dose Assessment Under NRC's Radiological Criteria for License Termination Rule, p. 13-27. In T.J. Nicholson and J.D. Parrott [ed.] Proceedingss of the Workshop on Review of Dose Modeling Methods for Demonstration of Compliance with the Radiological Criteria for License Termination. U.S. Nuclear Regulatory Commission NUREG/CP-0163, Washington, D.C.

Johnson, P.C., M.B. Hertz, and D.L. Byers. 1990. Estimates for hydrocarbon vapor emissions resulting from service station remediations and buried gasoline-contaminated soils, p. 295-326. In P.T. Kostecki and E. J. Calabrese [ed.] Petroleum Contaminated Soils, Vol. 3, Lewis Publishers, Chelsea, Mich.

Kersting, A.B., D.W. Efurd, D.L. Finnegan, D.J. Rokop, D.K. Smith, and J.L. Thompson. 1999 (January 7). Migration of plutonium in ground water at the Nevada Test Site. Nature 397:56-59.

National Research Council. 1983. Risk Assessment in the Federal Government: Managing the Process. Committee on the Institutional Means for Assessment of Risks to Public Health, National Academy Press, Washington, D.C. 191 pp.

National Research Council. 1994. Science and Judgement in Risk Assessment. Committee on Risk Assessment of Hazardous Air Pollutants, National Academy Press, Washington, D.C.

National Research Council. 1996. Understanding Risk; Informing Decisions in a Democratic Society. Committee on Risk Characterization, National Academy Press, Washington, D.C. 249 pp.

Regens, J.L., C. Travis, K.R. Obenshain, C. Whipple, J.T. Gunter, V. Miller, D. Hoel, G. Chieruzzi, M. Clauberg, and P.D. Wills. 1999. Multimedia Modeling and Risk Assessment. Medical University of South Carolina Press, Columbia, S.C.

Rust Geotech. 1996. Vadose Zone Monitoring Project at the Hanford Tank Farms: Tank Summary Data Report for Tank SX-112. Grand Junction Projects Office Report GJ-HAN-14, Tank SX112, Grand Junction, Colo.

Streile, G.P., K.D. Shields, J.L. Stroh, L.M. Bagaasen, G. Whelan, J.P. McDonald, J.G. Droppo, and J.W. Buck. 1996. The Multimedia Environmental Pollutant Assessment System (MEPAS): Source-Term Release Formulations. Pacific Northwest National Laboratory PNNL-11248, Richland, Wash.

U.S. Department of Energy. 1996 (June). Remedial Design Report/Remedial Action Work Plan for the 100 Area. Richland Office DOE/RL-96-17, Richland, Wash.

U.S. Department of Energy. 1998 (December 17). Groundwater/Vadose Zone Integration Project Specifications. Richland Operations Office DOE/RL-98-48, Draft C, Richland, Wash.

U.S. Department of Energy. 1999 (June). Groundwater/Vadose Zone Integration Project: Volume I-Summary Description; Volume II-Science and Technology Summary and Description; and Volume III-Background Information and State of Knowledge. Richland Operations Office DOE/RL-98-48, Rev.0, Richland, Wash.

U.S. Environmental Protection Agency. 1996. MMSOILS Model. Washington, D.C.

Whelan, G., J.P. McDonald, and C. Sato. 1996. Multimedia Environmental Pollutant Assessment System (MEPAS): Groundwater Pathway Formulations. Pacific Northwest National Laboratory PNNL-10907, Richland, Wash.

Winograd, I.J., and W. Thordarson. 1975. Hydrogeologic and Hydrochemical Framework, South-Central Great Basin, Nevada-California, with Special Reference to the Nevada Test Site. U.S. Geological Survey Professional Paper 712-C, Washington, D.C. 125 pp.

Yu, C., A.J. Zielen, J-J. Cheng, Y.C. Yuan, L.G. Jones, D.J. LePoire, Y.Y. Wang, C.O. Loureiro, E.K. Gnanapragasam, E. Faillace, A. Wallo III, W.A. Williams, and H. Peterson. 1993. Manual for Implementing Residual Radioactive Material Guidelines Using RESRAD Version 5.0. Argonne National Laboratory ANL/EAD/LD-2, Argonne, Ill.

Zaslavaky, D., and G. Sinai 1981. Surface hydrology: I-V. Journal of the Hydrology Division, American Society of Civil Engineers 107(HYI):1-93.

APPENDIX H

Biographical Sketches of Committee Members and Consultants

THOMAS M. LESCHINE, *Chair*, is associate professor in the School of Marine Affairs at the University of Washington. He is a former fellow in marine policy and a policy associate at the Woods Hole Oceanographic Institution. He is the chair of the National Research Council Committee on Remediation of Buried and Tank Wastes and also has served on the National Research Council Committee on Risk Assessment and Management of Marine Systems. His major research interest is in the area of environmental decision-making as it relates to marine environmental protection and the use of scientific and technical information in environmental decision-making. He is also interested in the use of mathematical modeling and systems analysis in environmental management. Dr. Leschine received his PhD in mathematics from the University of Pittsburgh.

MARY R. ENGLISH, *Vice Chair*, is a research leader for the Energy, Environment and Resources Center at the University of Tennessee, and a member of its Waste Management Research and Education Institute. She previously worked in environmental planning for state government and as a consultant. She was a member of the National Research Council Board on Radioactive Waste Management from 1995 through 1999. Dr. English received a BA in American Literature from Brown University, an MS in regional planning from the University of Massachusetts, and a PhD in sociology from the University of Tennessee.

DENISE BIERLEY is an independent environmental consultant specializing in environmental management, education, and policy issues. She has over 25 years of diverse experience including program management, natural resource management, radioactive and hazardous waste management, and regulatory compliance. She is currently working on salmon management issues in the Pacific Northwest. She holds BS degrees in biology and geology from Wright State University.

GREGORY R. CHOPPIN is the R.O. Lawton Distinguished Professor of Chemistry at Florida State University. Dr. Choppin's research includes nuclear chemistry, physical chemistry of the actinides and lanthanides, environmental behavior of actinides, chemistry of the f-Elements, separation science of the f-Elements, and concentrated electrolyte solutions. During a postdoctoral period at the Lawrence Radiation Laboratory, University of California, Berkeley, he participated in the discovery of mendelevium, element 101. His research activities have been recognized by the American Chemical Society's Award in Nuclear Chemistry and Southern Chemist Award, the Manufacturing Chemists award in Chemical Education, a Presidential Citation Award of the American Nuclear

Society, and the Chemical Pioneer Award of the American Institute of Chemistry. He has served on numerous National Research Council committees and currently is a member of the Board on Radioactive Waste Management. He received a BS degree in chemistry from Loyola University, New Orleans, a PhD in chemistry from the University of Texas, Austin, an honorary DTc from Chalmers University, Goteborg, Sweden, and an honorary DSc from Loyola University.

JAMES H. CLARKE is professor of the practice of civil and environmental engineering at Vanderbilt University. He has over 25 years of experience in environmental chemistry and chemical risk assessment. His primary areas of interest include environmental forensic science, the fate and transport of chemicals in the environment, the design of data acquisition programs for evaluation of the risks associated with chemical releases, and emerging technologies for hazardous waste site remediation. Dr. Clarke is a member of the American Academy of Forensic Sciences, the Tennessee Academy of Science, and the American Chemical Society. He received a BA in chemistry from Rockford College, Rockford, Illinois, and a PhD in theoretical physical chemistry from the Johns Hopkins University, Baltimore, Maryland.

ALLEN G. CROFF is associate director of the Chemical Technology Division at Oak Ridge National Laboratory (ORNL). His areas of focus include initiation and technical management of research and development involving waste management, national security, nuclear fuel cycles, transportation, energy efficiency, and renewable energy. Since joining ORNL in 1974, he has been involved in numerous technical studies that have focused on waste management and nuclear fuel cycles, including: (1) updating and implementing the ORIGEN-2 computer code; (2) developing a risk-based, generally applicable radioactive waste classification system; (3) multidisciplinary development and assessment of actinide partitioning and transmutation; and (4) leading and participating on multidisciplinary national and international technical committees. He has a BS in chemical engineering from Michigan State University, a nuclear engineer degree from the Massachusetts Institute of Technology, and an MBA from the University of Tennessee.

WILLIAM R. FREUDENBURG is a professor of rural sociology and environmental studies at the University of Wisconsin-Madison. He is a specialist on the human aspects of risk assessment and risk management, and has done extensive research on nuclear and other energy technologies. He has served as chair of Section K (Social, Economic and Political Sciences) of the American Association for the Advancement of Science. He has served on several NRC committees and federal advisory committees relating to energy and waste management issues. He was the first congressional fellow from the American Sociological Association to serve in the U.S. House of Representatives. Dr. Freudenburg received his PhD in sociology from Yale University in 1979.

DONALD R. GIBSON, JR. is the program manager for TRW's consolidated research and development contract at the Joint National Test Facility, which provides missile defense related analysis, system level engineering, integration, and test and evaluation support for the development, acquisition, and deployment of missile defense systems and architectures. Prior to this position he was deputy program manager and technical director for TRW's Joint Training, Analysis, and Simulation Center Support Team, manager of TRW's Systems Analysis and Integration Department supporting the Department of Energy's Office of Civilian Radioactive Waste Management, manager of the Survivability and Engineering Laboratory for TRW's Ballistic Missiles Division, and a design physicist for Los Alamos National Laboratory. Dr. Gibson holds a MS and PhD in nuclear engineering from the University of Illinois.

NAOMI H. HARLEY received a PhD in radiological physics in 1971, and a ME in nuclear engineering in 1967 from New York University. She also holds a BE in electrical engineering from the Cooper Union and an A.P.C. in management from the New York University Graduate Business School. Dr. Harley is currently a research professor of environmental medicine at the New York University School of Medicine, where she also serves on the Medical Isotopes Committee. She is a member in the National Council on Radiation Protection and Measurements, and an advisor to the U.S. Delegation of the United Nations Committee on the Effects of Atomic Radiation. Dr. Harley is a member of the Editorial Board of the journal *Environment International* and a fellow of the Health Physics Society. She has published over 100 journal articles, six book chapters, and she holds three patents at New York University for radiation detection devices. Her expertise is in radiation carcinogenesis, and her major research interests include measurement of inhaled or ingested radionuclides, the modeling of their fate within the human body, and the calculation of the detailed radiation dose to the cells specific to carcinogenesis.

JAMES H. JOHNSON, JR. is professor of civil engineering and dean of the College of Engineering, Architecture, and Computer Sciences at Howard University. Dr. Johnson's research interests have focused mainly on the reuse of wastewater treatment sludges and the treatment of hazardous substances. His recent research has included the refinement of composting technology for the treatment of contaminated soils, chemical oxidation and cometabolic transformation of explosive contaminated wastes, biodegradation of fuel-contaminated groundwater, the evaluation of environmental policy issues in relation to minorities and development of environmental curricula. Currently, he also serves as associate director of the Great Lakes and Mid-Atlantic Hazardous Substance Research Center, member of the Environmental Engineering Committee of U.S. EPA's Science Advisory Board, the National Research Council (NRC) Board on Radioactive Waste Management, and the NRC Committee on Remediation of Buried and Tank Wastes. Dr. Johnson received his BS from Howard University, MS from University of Illinois, and PhD from the University of Delaware. He is a fellow of the American Society of Civil Engineers, a registered professional engineer, and a diplomate of the American Academy of Environmental Engineers.

SHLOMO P. NEUMAN received a BS in geology from the Hebrew University in Jerusalem, Israel, and MS and PhD in engineering science from the University of California at Berkeley. Since 1975 he has served as professor, and since 1988 as Regents' Professor, of hydrology and water resources at the University of Arizona in Tucson. Prior to arriving in Tucson he was visiting associate professor of civil engineering at the University of California, Berkeley, and senior scientist at the Agricultural Research Organization at Bet-Dagan, Israel. He is fellow of the American Geophysical Union and the Geological Society of America, and member of the National Academy of Engineering, Sigma Xi, Society of Petroleum Engineers, American Association of Ground Water Scientists and Engineers, the International Association of Hydrogeologists, the American Institute of Hydrology, and the Arizona Hydrological Society. He holds numerous awards and has published over 200 articles, books, and reports.

W. HUGH O'RIORDAN is an attorney with Givens Pursley, LLP, in Boise, Idaho. He received a BA and JD from the University of Arizona and a LLM from George Washington University in environmental law. Since entering private practice in 1980, he has specialized in environmental, natural resources, and energy and administrative law on a state and federal level. He represents corporate and individual clients in matters involving environmental statutes. He is a member of the American Bar Association and a member of the Arizona, District of Columbia, and Idaho Bar Associations.

EDWIN WOODS ROEDDER received his BA from Lehigh University in 1941, his AM from Columbia University in 1947, and his PhD in geology in 1950. He also holds an honorary DSc, from Lehigh University (1976). Since 1987 Dr. Roedder has been an associate in the Department of Earth Planetary Sciences at Harvard University. From 1955 until 1987 he was employed by the U.S. Geological Survey as a geologist. Dr. Roedder served a member of the Committee on Geochemical Research at the National Science Foundation from 1954 until 1955. His honors and awards include an Exceptional Achievement Medal from NASA in 1973; the Werner Medal of the German Mineralogical Association, 1985; the Roebling Medal of the Mineralogical Society of America, 1986; and the Penrose medal of the Society of Economic Geologists, 1988. He is a member of the National Academy of Sciences; the Mineralogical Society of America (vice president from 1981-1982, president, 1982-1983); the American Geophysical Union; and Geochemical Society (president). His research interests are in the fields of ore deposition, fluid inclusions in minerals, studies of lunar materials, nuclear waste storage problems, and volcanology.

BENJAMIN ROSS is president of Disposal Safety, Incorporated, a consulting firm in Washington, D.C., which specializes in analysis of groundwater and soil contamination by hazardous radioactive and chemical waste. Dr. Ross also heads European Analytical Services, Inc., which represents Russian institutes selling technical services and products in the United States. Before starting Disposal Safety, Dr. Ross was a senior research scientist at GeoTrans, Inc., and a risk analyst with the Analytic Sciences Corporation. Dr. Ross received his AB in physics from Harvard University and his PhD in physics from the Massachusetts Institute of Technology.

RAYMOND G. WYMER is a retired director of the Chemical Technology Division of Oak Ridge National Laboratory. He is a specialist in radiochemical separations technology for radioactive waste management and nuclear fuel reprocessing. He is a member of the Advisory Committee on Nuclear Waste for the Nuclear Regulatory Commission. He is a consultant for the Oak Ridge National Laboratory and for the U.S. Department of Energy in the area of chemical separations technology. He consults for the U.S. Department of State and the U.S. Department of Energy on matters of nuclear nonproliferation. He is a fellow of the American Nuclear Society and the American Institute of Chemists, and has received the American Institute of Chemical Engineers' Robert E. Wilson Award in Nuclear Chemical Engineering and the American Nuclear Society's Special Award for Outstanding Work on the Nuclear Fuel Cycle. He received a BA from Memphis State University and an MA and PhD from Vanderbilt University.

CONSULTANTS

ROBERT M. BERNERO received his BA degree from St. Mary of the Lake (Illinois), a BS degree from the University of Illinois, and his MS degree from Rensselaer Polytechnic Institute. He has recently retired from 23 years of service with the U.S. Nuclear Regulatory Commission (USNRC), where he held numerous positions up to director of the Office of Nuclear Material Safety and Safeguards. Prior to joining the USNRC, he worked for the General Electric Company in nuclear technology for 13 years. He currently consults on nuclear safety-related matters, and served as a member of the Commission of Inquiry for an International Review of Swedish Nuclear Regulatory Activities in 1995 and 1996. His areas of interest include licensing, inspection, and environmental review of uses of nuclear technology and radioactive waste management.

ELIZABETH K. HOCKING received her JD from the Washington College of Law of the American University. Since 1989 Ms. Hocking has been a policy analyst with the Environmental Assessment Division of Argonne National Laboratory and is manager of its Environmental Policy Analysis section. Her research interests include federal property transfer, institutional controls, and statutory and policy changes to environmental remediation programs. She served as a note and comment editor on the *American University Administrative Law Journal* from 1990 to 1991.

APPENDIX I

Definitions of Terms Used in This Report

The committee uses certain terms throughout this report. Their definitions are assembled here to assist the reader.

Long-Term Institutional Management—A comprehensive approach to planning and decision-making for management of contaminated sites, facilities, and materials.

Contaminant Reduction—Activities that decrease the volume or toxicity of contaminants at a particular location. These include destruction, decontamination, treatment and processing, natural and radioactive decay, and removal.

Contaminant Isolation—Use of natural or engineered barriers and stabilization techniques to prevent or limit the migration of contaminants and to prevent human intrusion.

Contaminant Remediation—Contaminant reduction and contaminant isolation.

Stewardship—Activities that will be required to manage potentially harmful residual contamination left on site after cessation of remediation efforts, including:
- maintaining contaminant isolation and measures to monitor the migration and attenuation or evolution of residual contaminants;
- institutional controls (see definition below);
- conducting oversight and, if necessary, enforcement;
- gathering, storing, and retrieving information about residual contaminants and conditions on site, as well as about changing off-site conditions that may affect or be affected by residual contaminants;
- disseminating information about the site and related use restrictions;
- periodically reevaluating how well the total protective system is working;
- evaluating of new technological options to reduce or eliminate residual contaminants or to monitor and prevent migration of isolated contaminants; and
- supporting research and development aimed at improving basic understanding of both the physical and sociopolitical character of site environments and the fate, transport, and effects of residual site contaminants.

Institutional Controls—Restrictions on land access or use through such devices as easements, deed notification, zoning, permits, fences, signs, government ownership, and leases; also, legal measures to ensure continued access to privatized sites for the purpose of monitoring and, if necessary, further remediation.

Contextual Factors—Factors that can affect the nature and extent of the measures taken under long-term institutional management; seven factors in particular often constrain the range of decisions and actions realistically available:
- risk;
- scientific and technical capability;
- institutional capability;
- cost;
- laws and regulations;
- values of interested and affected parties; and
- other sites.